Glia in Health and Disease

Edited by Tania Spohr

Published in London, United Kingdom

IntechOpen

Supporting open minds since 2005

Glia in Health and Disease
http://dx.doi.org/10.5772/intechopen.82593
Edited by Tania Spohr

Contributors
Juan Andres Orellana, Soraya L. Valles, Adrian Jorda, Antonio Iradi, Martin Aldasoro, Federico Burguet, Venkat Venkataraman, Anuradha Krishnan, Hao Wu, Michal Izrael, Shalom Guy Slutsky, Michel Revel, Juliana De Mattos Coelho-Aguiar, Carla Pires Veríssimo, Deiziane Viana Da Silva Costa, Ana Lucia Gomes, Beatriz Thomasi, Ana Carina Bon Frauches, Gerly Anne De Castro Brito, Vivaldo Moura- Neto, Fabiana Pereira Ribeiro, Allison Soung, Robyn Klein, Ana Karolina Nunes, Haniel Fernandes, Sura Rocha

First published in London, United Kingdom, 2020 by IntechOpen
IntechOpen is the global imprint of INTECHOPEN LIMITED, registered in England and Wales, registration number: 11086078, 7th floor, 10 Lower Thames Street, London, EC3R 6AF, United Kingdom
Printed in Croatia

British Library Cataloguing-in-Publication Data
A catalogue record for this book is available from the British Library

Additional hard and PDF copies can be obtained from orders@intechopen.com

Glia in Health and Disease
Edited by Tania Spohr
p. cm.
Print ISBN 978-1-78985-253-0
Online ISBN 978-1-78985-254-7
eBook (PDF) ISBN 978-1-78985-614-9

For EU product safety concerns:
IN TECH d.o.o., Prolaz Marije Krucifikse Kozulić 3, 51000 Rijeka, Croatia, info@intechopen.com or visit our website at intechopen.com.

Meet the editor

Dr. Tania Spohr finished her PhD in Morphological Sciences at the Federal University of Rio de Janeiro, when she was honored with the prize FAPERJ 30 Years because her PhD thesis was considered the best thesis in her field in the state of Rio de Janeiro in 2009. Her PhD thesis also won the Carlos Chagas Filho Award because it was considered the best PhD thesis of the Morphological Sciences Graduate Program in UFRJ in that year. Presently, she is a research associate scientist at the Instituto Estadual do Cérebro Paulo Niemeyer in Rio de Janeiro. Her research interest covers the areas of glial cells (healthy and pathological). She has more than 20 research articles published and five book chapters.

Contents

Preface XI

Section 1
Neuroscience 1

Chapter 1 3
Synaptic Functions of Astroglial Hemichannels
by Juan A. Orellana

Chapter 2 15
Astrocytic S100B, Blood-Brain Barrier and Neurodegenerative Diseases
by Anuradha Krishnan, Hao Wu and Venkat Venkataraman

Chapter 3 35
The Enteric Glial Network Acts in the Maintenance of Intestinal Homeostasis and in Intestinal Disorders
by Juliana de Mattos Coelho-Aguiar, Carla Pires Veríssimo, Deiziane Viana da Silva Costa, Beatriz Bastos de Moraes Thomasi, Ana Carina Bon Frauches, Fabiana Pereira Ribeiro, Ana Lucia Tavares Gomes, Gerly Anne de Castro Brito and Vivaldo Moura-Neto

Chapter 4 63
Involvement of Astrocytes in the Process of Metabolic Syndrome
by Ana Karolina Santana Nunes, Haniel Soares Fernandes and Sura Wanessa Santos Rocha

Section 2
Neuropathology 79

Chapter 5 81
Astrocytes and Inflammatory Processes in Alzheimer's Disease
by Soraya L. Valles, Federico Burguet, Antonio Iradi, Martin Aldasoro, Jose M. Vila, Constanza Aldasoro and Adrián Jordá

Chapter 6 109
Astrocytes: Initiators of and Responders to Inflammation
by Allison Soung and Robyn S. Klein

Chapter 7 131
Astrocytes in Pathogenesis of Multiple Sclerosis and Potential Translation into Clinic
by Izrael Michal, Slutsky Shalom Guy and Revel Michel

Preface

It was the German pathologist Rudolf Virchow who described glial cells for the first time. In 1846, he analyzed postmortem human tissues, and considered those cells to be merely a "glue" having only a passive supporting role. Nowadays, glial cells are considered more than passive components and have become active partners with neurons. Thus, they seem to be much more actively involved in brain function than was formerly thought. Glial cells comprise microglia, oligodendrocytes, and astrocytes in the central nervous system, and Schwann cells, satellite cells, and enteric glial cells in the peripheral nervous system. It is already well established that neuron–glia interactions control several processes of brain development such as neurogenesis, myelination, synapse formation, neuronal migration, proliferation, differentiation, and even neuronal signaling. In all these processes astrocytes have important roles: neurotransmitter clearance, ion buffering, and neuronal trophic support by secreting members of the epidermal growth factor family, transforming growth factor, neuregulin, fibroblast growth factor, nerve growth factor, and ciliary neurotrophic factor, for instance. Moreover, astrocytes are also involved in other functions, such as synapse development, blood–brain barrier formation, and neurogenesis.

Despite all the roles of glial cells in a healthy nervous system, they are also involved in neurological disorders or diseases. In response to injury and diseases, glial cells suffer a process termed astrogliosis that induces proliferation, progressive cell hypertrophy, progressive alteration in molecular expression, and scar formation. Studies have demonstrated that the malfunction of glial cells plays a pivotal role in several neurodegenerative diseases, such as amyotrophic lateral sclerosis, Parkinson's disease, Alzheimer's disease (AD), and multiple sclerosis (MS). During the development of several neurological diseases, there is an increase in the inflammation process that is related to the progression and worsening of the symptoms.

In this book, we will highlight the role played by glial cells in the central and peripheral nervous systems in healthy and unhealthy individuals by giving particular attention to the enteric nervous system (ENS). Among all processes involved, we will specifically discuss the importance of ENS in the control of gut homeostasis, in the interaction with the immune system, and its participation in pathological conditions such as metabolic syndrome.

In particular, the relevance of astrocytes will be explored during synaptic transmission and regulation of plasticity by releasing gliotransmitters. Ultimately, we will highlight the influence of astrocytes during the development of a number of neurodegenerative diseases, such as MS and AD. We will focus on how the serum levels of the astrocytic protein S100B can be used as a biomarker for clinical decisions for the onset and progression of neurodegenerative diseases.

Tania Spohr
Instituto Estadual do Cérebro Paulo Niemeyer,
Rio de Janeiro, Brazil

Section 1

Neuroscience

Chapter 1

Synaptic Functions of Astroglial Hemichannels

Juan A. Orellana

Abstract

In recent decades, astrocytes have gained ground in their protagonist role at the synapses, challenging the old-historic idea that neurons are the unique functional units in the nervous system. Although for a long time considered merely supportive elements, astrocytes are now recognized as a source of gliotransmitter release that regulates synaptic transmission and plasticity. Despite the initial evidence that supported gliotransmission depends on intracellular Ca^{2+}-mediated vesicular release, recent data indicate that hemichannels may constitute an alternative non-vesicular route for gliotransmitter efflux. These channels are plasma membrane channels formed by the oligomerization of six connexins around a central pore. Hemichannels are permeable to ions and signaling molecules—such as ATP, glutamate, and Ca^{2+}—constituting a pathway of diffusional interchange between the cytoplasm and the extracellular milieu. Connexin 43 is the main hemichannel-forming protein in astrocytes and is highly regulated under physiological and pathological conditions. In this chapter, the available data supporting the idea that hemichannels are chief components in tuning the synaptic gain in either resting or stimulated conditions is discussed.

Keywords: connexin 43, astrocyte, gliotransmission, brain, neuron

1. Introduction

In order to ensure a proper response to external stimuli, organisms have created complex and coordinated neural structures that allow the sophisticated analysis of information. As the central nervous system (CNS) evolved from a basic network structure to compacted ganglia and centralized brains, two types of connections emerged as specialized structures favoring the integration of neural networks [1]. In 1897, Sherrington proposed the point of functional contact between neurons as the specific area at which transfer of information takes place and named it "synapsis," soon shortened to the "synapse," from the Greek word *sunáptō* (to clasp) [2]. This specialized structure is known today as the chemical synapse and transfers electrical information unidirectionally from presynaptic to postsynaptic neurons through the release of neurotransmitters, which, acting upon postsynaptic receptors, initiate a second electrical signal [1]. In the late 1950s, Furshpan and Potter reported a series of experiments revealing that synaptic transmission in the crayfish is bidirectional and voltage-dependent, two properties substantially out of range of the criteria established for chemical transmission [3]. This study revealed the pioneer evidence in favor of the existence of electrical synaptic transmission. Unlike chemical synapse, the electrical synapse permits the bidirectional flow of ions between coupled neurons that come markedly close at intercellular specializations called gap junctions [4] (**Figure 1**). Nowadays, a growing body of evidence indicates that both mechanisms of synaptic transmission—chemical and electrical—are complementary and highly intermodulated to ensure proper brain development and function [1].

The traditional notion of neurons being the only functional elements in the synapse has been questioned with the finding that intracellular Ca^{2+} ($[Ca^{2+}]_i$) waves within and among astrocytes underlie the regenerative (nondissipative) transfer of biological signals [5–7]. Although astrocytes are not electrically silent cells [8], $[Ca^{2+}]_i$ signals are their principal fast time-scale mechanism for allowing intra- and intercellular signaling [9]. These signals base their origin on the extracellular influx of Ca^{2+} via ion channels and through Ca^{2+} release from intracellular stores, resulting in $[Ca^{2+}]_i$ transients that differ in frequency, kinetics, and spatial spread depending on the astroglial anatomical region [10]. Endowed with this machinery and along with pre- and postsynaptic neuronal elements, astrocytes embrace the "tripartite synapse"—the Rosetta stone of the chemical synaptic transmission—in where they sense neurotransmission and respond to it by releasing biomolecules that regulate neuronal activity called "gliotransmitters" (i.e., glutamate, D-serine, and ATP) [11]. Intracellular $[Ca^{2+}]_i$ waves can spread among astrocytes to finally reach the terminal processes or "endfeet" of specialized astrocytes that contact the endothelium [12]. There, vasoactive molecules are released, permitting astrocytes to modulate the cerebral blood flow (CBF) and delivery of energy substances (i.e., glucose and lactate) with potentially significant consequences for neuronal firing and higher brain functions [13]. Indeed, a single astrocyte may contact over 100,000 synapses in rodents and up to 2,000,000 synapses in humans, revealing that they actually form a syncytium with multiple connections [14].

Nowadays, diverse mechanisms have been proposed to lead to gliotransmitter release (**Figure 1**), including Ca^{2+}-dependent exocytosis [15–17], carrier membrane transport [18], and opening of a wide range of channels. Among the latter group, volume-regulated anion channels [19–21], P2X7 receptors [22–24], Ca^{2+}-dependent Cl^- channel bestrophin 1 [25, 26], and hemichannels [27–30] are included. This

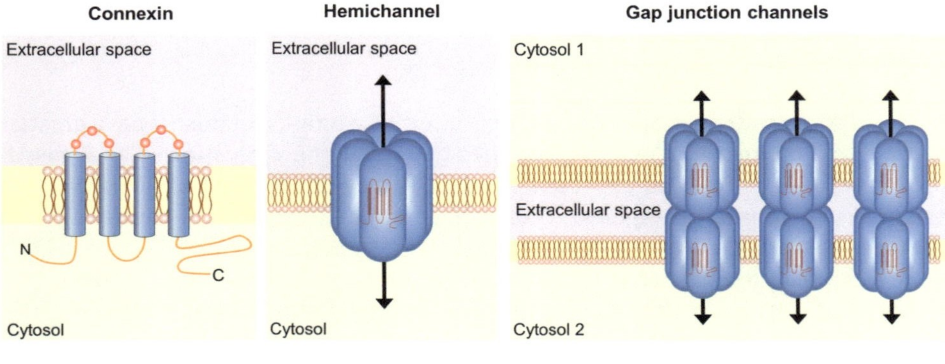

Figure 1.
Basic structure of connexin-based channels. Connexins have four α-helical transmembrane domains connected by two extracellular loops and one cytoplasmic loop; both the amino- and carboxy-termini are intracellular. The relative positions of the extracellular loop cysteines (red balls) are also shown. Hemichannels (also known as connexons) are formed by the oligomerization of six subunit connexins around a central pore. Under resting conditions, hemichannels remain preferentially closed, but they may be activated by diverse physiological and pathological conditions and offer a diffuse transmembrane route between the intra- and extracellular milieu. Hemichannels dock each other to form functional cell-to-cell channels termed gap junction channels (right panel). Gap junction channels aggregate in well-known anatomical structures called gap junctions to facilitate the intercellular cytoplasmic exchange of metabolites, second messengers, and ions.

chapter reviews and discusses recent data supporting a role for hemichannels as pathways for gliotransmission and relevant actors in that tuning of synaptic transmission and plasticity.

2. Structure and major functions of hemichannels

During the past decade, a growing body of evidence began to support a novel mechanism of autocrine/paracrine communication underlying gliotransmission and astrocyte-to-neuron communication: hemichannel-mediated signaling [31]. Each hemichannel is composed of the oligomerization of six protein subunits called connexins around a central pore (**Figure 1**). Connexins embrace a highly conserved protein family encoded by 21 genes in humans and 20 in mice, with orthologs in other vertebrate species [32]. These proteins are abundantly expressed in brain cells [33], including astrocytes [34], and they are named after their predicted molecular mass expressed in kDa, for instance, connexin 43 (Cx43) has a molecular mass of ~43 kDa [35]. For several years, the key function attributed to hemichannels was to constitute the building blocks of the gap junction channels, which are intercellular channels that allow the direct cytoplasmic exchange between contacting cells [35]. Nonetheless, in the 1990s, pioneering findings by Paul and colleagues revealed the presence of functional and solitary hemichannels in "nonjunctional" membranes [36]. Today, it is well accepted that these channels act like aqueous pores, providing a diffusional route of exchange for ions and molecules between the intra- and extracellular space [37]. Across the different tissues, hemichannels allow the cellular release of relevant quantities of autocrine and paracrine signaling molecules (e.g., ATP, glutamate, D-serine, NAD^+, and PGE_2), as well as the influx of other substances (i.e., Ca^{2+} and glucose) [37].

Since their discovery, hemichannels have been linked with cellular damage. This idea came from early studies suggesting that osmotic and ionic imbalances induced by the uncontrolled influx of Na^{2+} and Cl^- through hemichannels could result in further cell swelling and plasma membrane breakdown [36]. In addition, it has been proposed that because hemichannels are permeable to Ca^{2+}, their uncontrolled opening could lead to Ca^{2+} overload and the consequent production of free radicals, lipid peroxidation, and plasma membrane damage [38]. Alternatively, exacerbated hemichannel activity could also induce the release of molecules that at high concentration may be toxic for neighboring cells, such as glutamate, in the case of the CNS [39]. Despite the above, in the last decade, a substantial body of studies has proposed that hemichannels may underpin pivotal neurophysiological functions, such as synaptic efficacy, neural activity, signal processing, cognition, and behavior [27, 28, 40–44].

3. Astroglial hemichannels and their role in synaptic transmission and plasticity

Although rat, mouse, and human astrocytes express abundantly Cx30 and Cx43, as well as Cx26 [45–49], at the moment, Cx43 is the only connexin probed to form functional hemichannels in astrocytes [50]. The opening of astroglial Cx43 hemichannels has been linked with the release of different gliotransmitters (e.g., glutamate, ATP, D-serine, lactate), as well as with the influx of extracellular Ca^{2+} and glucose. Seminal studies by Torres and colleagues demonstrated for the first time that astrocyte hemichannels may act as both sensors and modulators of synaptic activity [43]. Using UV-photolysis of caged MNI-glutamate in hippocampal slices, they found that specific deletion of Cx43 abrogates ATP-dependent spreading of slow Ca^{2+} waves among astrocytes. Furthermore, these slow Ca^{2+} waves were potentiated when authors used slices from transgenic mice with an astrocyte-targeted point mutation ($Cx43^{G138R}$) that leads to an increased Cx43 hemichannel opening [51]. In addition, they observed that depolarization of inhibitory interneurons from the stratum radiatum reduced CA1 excitatory transmission via the astroglial Cx43 hemichannel-mediated release of ATP and subsequent stimulation of interneuronal P2Y1 receptors [43]. These data shed light for the first time about how astrocyte Cx43 hemichannels may underpin a negative feedback mechanism elicited during sustained excitation to prevent excitotoxicity (**Figure 2**).

Although in normal astrocytes few Cx43 hemichannels are in the plasma membrane and most of them with a low open probability, recent findings have described that they facilitate the release of ATP under basal conditions [27, 41]. Chever and co-workers observed that basal release of ATP via astroglial Cx43 hemichannels is enough to boost the CA1 synaptic transmission triggered by stimulation of Schaffer collaterals, an effect mediated by purinergic receptors [27] (**Figure 2**). Likely the insertion of postsynaptic AMPA receptors as a result of the activation of P2X7 receptors could explain the ATP-dependent potentiation of glutamatergic transmission, as reported before in other brain areas [52]. Astroglial hemichannels also have been found to regulate neuronal activity in the olfactory bulb (OB) [41]. There, the group of Giaume demonstrated that pharmacological inhibition of Cx43 hemichannels decreased the firing and amplitude of depolarized states in mitral cells. Similar findings were observed in mitral cells of OB slices with specific astroglial deletion of Cx43 [41] or in slices treated with A1 adenosine receptor antagonists. These findings denote that likely astrocyte Cx43 hemichannels enhance the amplitude of depolarized states of mitral cells through the release of ATP and its further breakdown to

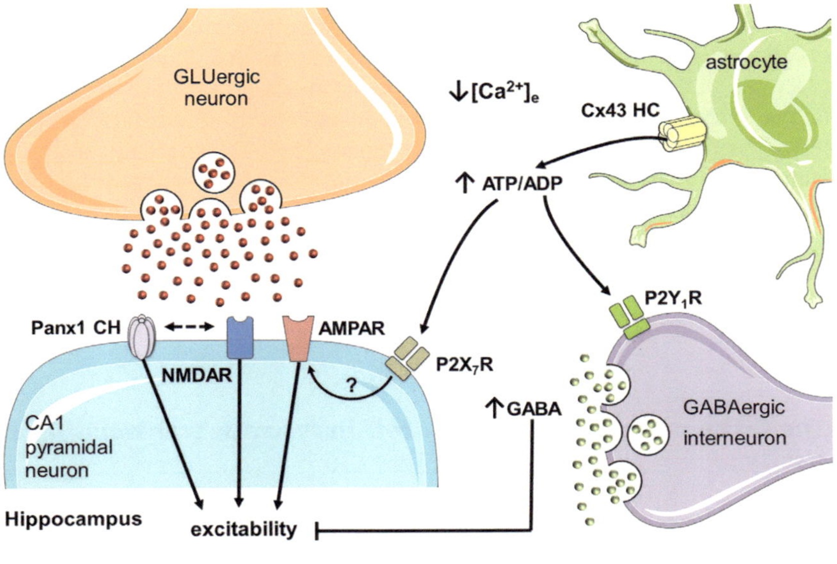

Figure 2.
Possible role of astroglial hemichannels in hippocampal synaptic transmission. During the basal firing of glutamatergic neurons in the hippocampus, Ca^{2+} influx into neurons results in a localized reduction in $[Ca^{2+}]_e$, which in turn opens Cx43 hemichannels (HCs) on astrocytes [43]. The latter lead to the release of ATP, being this crucial for sustaining basal excitatory synaptic transmission [27]. Likely this phenomenon takes place via the activation of P2X7 receptors and further insertion of AMPA receptors in postsynaptic terminals [52]. Alternatively, the conversion of ATP to ADP could depolarize and augment firing in interneurons via P2Y1 receptors, therefore, enhancing inhibitory transmission [43].

adenosine (**Figure 3**). Usually, A1 receptors induce the presynaptic inhibition of glutamate release, reduced postsynaptic NMDAR activation, and decreased Ca^{2+} influx [53]. Therefore, it is possible that the adenosine-mediated enhancement of depolarized states is due to the suppression of inhibitory juxtaglomerular interneurons, as occurred in other brain areas [54] (**Figure 3**).

Astroglial Cx43 hemichannels have been involved not only in synaptic function and transmission but also in synaptic plasticity. High-frequency stimulation (HFS)

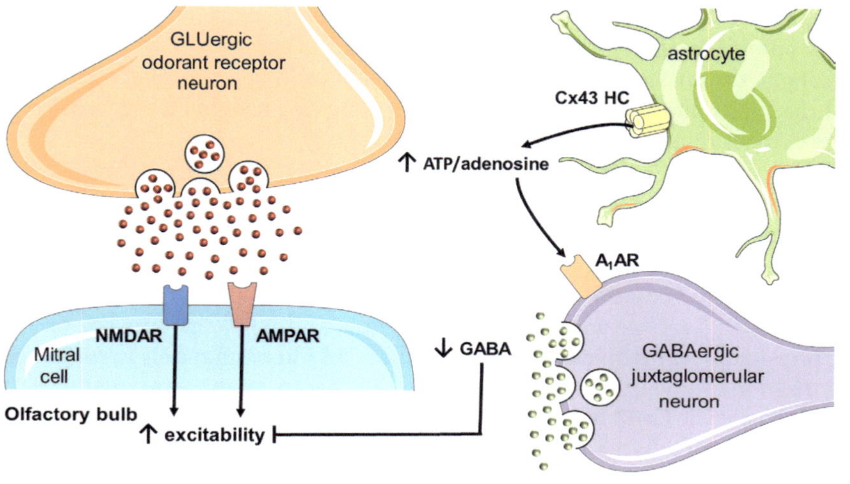

Figure 3.
Implications of astroglial hemichannel activity in neuronal oscillations of the olfactory bulb. Spontaneous neuronal activity in the glomerular layer of the olfactory bulb requires glutamatergic transmission. In this scenario, astrocytes display a basal release of ATP via Cx43 hemichannels (HCs) [41]. The adenosine derived from ATP may reduce the activity of GABAergic inhibitory juxtaglomerular neurons through the stimulation of A1 adenosine receptors. This permits the basal slow oscillations of up and down states of mitral cells in the olfactory bulb.

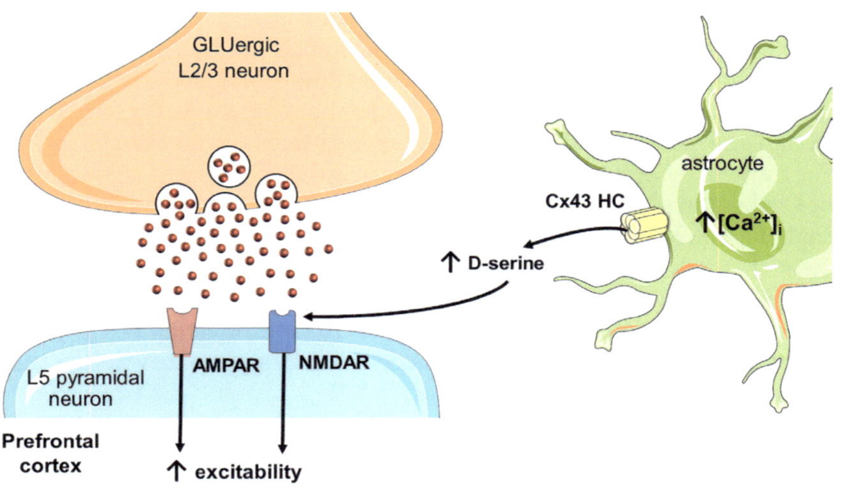

Figure 4.
Astroglial hemichannels and their impact on synaptic plasticity in the prefrontal cortex. In the prefrontal cortex, continuous stimulation of layer 2/3 neurons induces long-term potentiation (LTP) of NMDA and AMPA receptor currents in layer 5 pyramidal neurons. In this context, $[Ca^{2+}]_i$ is needed for the opening of Cx43 hemichannels (HCs) in astrocytes [28], which cause release of D-serine. This gliotransmitter facilitates LTP of NMDA and AMPA excitatory synaptic currents mediated by high-frequency stimulation.

of neuronal layer 2/3 (L2/3) triggers glutamatergic synaptic transmission in pyramidal cells at layer 5 (L5) of the prefrontal cortex (PFC) [55]. In this context and using PFC slices, Meunier and colleagues observed that genetic ablation of Cx43 or inhibition of Cx43 hemichannels strongly counteracts the NMDAR-dependent excitatory postsynaptic currents (EPSCs) and increases AMPA/NMDA current ratio induced by HSF in L5 [28]. Relevantly, the latter responses did not occur when D-serine was added at the recording media, revealing that the release of D-serine and astroglial hemichannel function are linked and modulate NMDAR-dependent synaptic transmission in PFC pyramidal cells. Furthermore, when $[Ca^{2+}]_i$ was clamped or D-serine production was inhibited in the L5 astroglial network, HFS failed to potentiate the NMDAR-dependent synaptic currents [28] (**Figure 4**). Accordingly, the authors hypothesized that potentiation of glutamatergic transmission at the PFC relies on $[Ca^{2+}]_i$-mediated opening of astroglial Cx43 hemichannels and the further release of D-serine (**Figure 4**).

The impact of astroglial hemichannels on synaptic transmission and plasticity has a subsequent echo on higher brain function and behavior. Indeed, in vivo inhibition of Cx43 hemichannels at the basolateral amygdala causes transitory and specific amnesia for auditory fear conditioning [42]. Remarkably, learning capacity was restored by the co-administration of a cocktail of supposed gliotransmitters (lactate, glutamate, D-serine, glutamine, glycine, and ATP), evidencing for the first time a physiological involvement for astroglial Cx43 hemichannels in higher brain function. In the same line, a recent study found that intraventricular administration of Gap19, a specific Cx43 hemichannel blocker [56], significantly impairs the spatial short-term memory, as assayed with the delayed spontaneous alternation Y maze task [44].

4. Conclusions

The impact of functional astroglial hemichannels in synaptic transmission and plasticity may depend on the number of channels available in the plasma membrane, their open probability, and their conductance and/or selectivity. Of particular relevance is to disentangle how synaptic function is modulated by regulations in gating properties of astroglial hemichannels, as well as changes in their trafficking or de novo synthesis. Elucidating the latter will allow us to understand whether hemichannel opening in astrocytes tunes the temporal outcome for sculpting either short-term (milliseconds to a few minutes) or long-term (minutes to hours) plasticity in the nervous system. One point of concern is the urgent need of developing new molecular and pharmacological tools to specifically dissect the contribution of astroglial hemichannels to the function of neural networks without affecting other hemichannel-forming proteins in other brain cells (e.g., microglia, oligodendrocytes, and endothelial cells). Finally, although growing evidence in ex vivo preparations has extended our knowledge about the role of astroglial hemichannels in gliotransmission, additional data are necessary to demonstrate whether this truly occurs in vivo.

Acknowledgements

This work was supported by (i) the Comisión Nacional de Investigación Científica y Tecnológica (CONICYT) and Fondo Nacional de Desarrollo Científico y Tecnológico (FONDECYT) Grant 1160710 (to JAO) and (ii) the CONICYT and Programa de Investigación Asociativa (PIA) Grant Anillo de Ciencia y Tecnología

ACT1411 (to JAO). The author did part of the schematics with support of the free online Servier Medical Art repository (https://smart.servier.com/).

Conflict of interest

The author declares no conflict of interest.

Author details

Juan A. Orellana
Departamento de Neurología, Escuela de Medicina and Centro Interdisciplinario de Neurociencias, Facultad de Medicina, Pontificia Universidad Católica de Chile, Santiago, Chile

*Address all correspondence to: jaorella@uc.cl

References

[1] Pereda AE. Electrical synapses and their functional interactions with chemical synapses. Nature Reviews. Neuroscience. 2014;**15**(4):250-263. Epub 2014/03/13

[2] Foster M, Sherrington CS. Central nervous system. In: A Textbook of Physiology. London: Macmillan; 1897. p. 930

[3] Furshpan EJ, Potter DD. Transmission at the giant motor synapses of the crayfish. The Journal of Physiology. 1959;**145**(2):289-325. Epub 1959/03/03.

[4] Bennett MV, Zukin RS. Electrical coupling and neuronal synchronization in the mammalian brain. Neuron. 2004;**41**(4):495-511. Epub 2004/02/26

[5] Cornell-Bell AH, Finkbeiner SM, Cooper MS, Smith SJ. Glutamate induces calcium waves in cultured astrocytes: Long-range glial signaling. Science. 1990;**247**(4941):470-473

[6] Nedergaard M. Direct signaling from astrocytes to neurons in cultures of mammalian brain cells. Science. 1994;**263**(5154):1768-1771

[7] Parpura V, Basarsky TA, Liu F, Jeftinija K, Jeftinija S, Haydon PG. Glutamate-mediated astrocyte-neuron signalling. Nature. 1994;**369**(6483):744-747

[8] Dallerac G, Chever O, Rouach N. How do astrocytes shape synaptic transmission? Insights from electrophysiology. Frontiers in Cellular Neuroscience. 2013;**7**:159

[9] Rusakov DA. Disentangling calcium-driven astrocyte physiology. Nature Reviews. Neuroscience. 2015;**16**(4):226-233

[10] Bazargani N, Attwell D. Astrocyte calcium signaling: The third wave. Nature Neuroscience. 2016;**19**(2):182-189

[11] Perea G, Navarrete M, Araque A. Tripartite synapses: Astrocytes process and control synaptic information. Trends in Neurosciences. 2009;**32**(8):421-431. Epub 2009/07/21

[12] Allaman I, Belanger M, Magistretti PJ. Astrocyte-neuron metabolic relationships: For better and for worse. Trends in Neurosciences. 2011;**34**(2):76-87

[13] Weber B, Barros LF. The astrocyte: Powerhouse and recycling center. Cold Spring Harbor Perspectives in Biology. 2015;**7**(12):a020396

[14] Oberheim NA, Takano T, Han X, He W, Lin JH, Wang F, et al. Uniquely hominid features of adult human astrocytes. The Journal of Neuroscience. 2009;**29**(10):3276-3287. Epub 2009/03/13

[15] Bezzi P, Gundersen V, Galbete JL, Seifert G, Steinhauser C, Pilati E, et al. Astrocytes contain a vesicular compartment that is competent for regulated exocytosis of glutamate. Nature Neuroscience. 2004;**7**(6):613-620. Epub 2004/05/25

[16] Zhang Z, Chen G, Zhou W, Song A, Xu T, Luo Q, et al. Regulated ATP release from astrocytes through lysosome exocytosis. Nature Cell Biology. 2007;**9**(8):945-953. Epub 2007/07/10

[17] Imura Y, Morizawa Y, Komatsu R, Shibata K, Shinozaki Y, Kasai H, et al. Microglia release ATP by exocytosis. Glia. 2013;**61**(8):1320-1330. Epub 2013/07/05

[18] Rossi DJ, Oshima T, Attwell D. Glutamate release in severe brain ischaemia is mainly by reversed uptake.

Nature. 2000;**403**(6767):316-321. Epub 2000/02/05

[19] Kimelberg HK, Goderie SK, Higman S, Pang S, Waniewski RA. Swelling-induced release of glutamate, aspartate, and taurine from astrocyte cultures. The Journal of Neuroscience. 1990;**10**(5):1583-1591. Epub 1990/05/01

[20] Takano T, Kang J, Jaiswal JK, Simon SM, Lin JH, Yu Y, et al. Receptor-mediated glutamate release from volume sensitive channels in astrocytes. Proceedings of the National Academy of Sciences of the United States of America. 2005;**102**(45):16466-16471. Epub 2005/10/29

[21] Rudkouskaya A, Chernoguz A, Haskew-Layton RE, Mongin AA. Two conventional protein kinase C isoforms, alpha and beta I, are involved in the ATP-induced activation of volume-regulated anion channel and glutamate release in cultured astrocytes. Journal of Neurochemistry. 2008;**105**(6): 2260-2270. Epub 2008/03/05

[22] Duan S, Anderson CM, Keung EC, Chen Y, Swanson RA. P2X7 receptor-mediated release of excitatory amino acids from astrocytes. The Journal of Neuroscience. 2003;**23**(4):1320-1328. Epub 2003/02/25

[23] Suadicani SO, Brosnan CF, Scemes E. P2X7 receptors mediate ATP release and amplification of astrocytic intercellular Ca^{2+} signaling. The Journal of Neuroscience. 2006;**26**(5):1378-1385. Epub 2006/02/03

[24] Hamilton N, Vayro S, Kirchhoff F, Verkhratsky A, Robbins J, Gorecki DC, et al. Mechanisms of ATP- and glutamate-mediated calcium signaling in white matter astrocytes. Glia. 2008;**56**(7):734-749. Epub 2008/02/23

[25] Lee S, Yoon BE, Berglund K, Oh SJ, Park H, Shin HS, et al. Channel-mediated tonic GABA release from glia. Science. 2010;**330**(6005):790-796. Epub 2010/10/12

[26] Woo DH, Han KS, Shim JW, Yoon BE, Kim E, Bae JY, et al. TREK-1 and best1 channels mediate fast and slow glutamate release in astrocytes upon GPCR activation. Cell. 2012;**151**(1):25-40

[27] Chever O, Lee CY, Rouach N. Astroglial connexin43 hemichannels tune basal excitatory synaptic transmission. The Journal of Neuroscience. 2014;**34**(34):11228-11232. Epub 2014/08/22

[28] Meunier C, Wang N, Yi C, Dallerac G, Ezan P, Koulakoff A, et al. Contribution of astroglial Cx43 hemichannels to the modulation of glutamatergic currents by D-serine in the mouse prefrontal cortex. The Journal of Neuroscience. 2017;**37**(37):9064-9075. Epub 2017/08/20

[29] Stout CE, Costantin JL, Naus CC, Charles AC. Intercellular calcium signaling in astrocytes via ATP release through connexin hemichannels. The Journal of Biological Chemistry. 2002;**277**(12):10482-10488. Epub 2002/01/14

[30] Ye ZC, Wyeth MS, Baltan-Tekkok S, Ransom BR. Functional hemichannels in astrocytes: A novel mechanism of glutamate release. The Journal of Neuroscience. 2003;**23**(9):3588-3596. Epub 2003/05/09

[31] Abudara V, Retamal MA, Del Rio R, Orellana JA. Synaptic functions of hemichannels and pannexons: A double-edged sword. Frontiers in Molecular Neuroscience. 2018;**11**:435

[32] Leybaert L, Lampe PD, Dhein S, Kwak BR, Ferdinandy P, Beyer EC, et al. Connexins in cardiovascular and neurovascular health and disease: Pharmacological implications.

Pharmacological Reviews. 2017;**69**(4):396-478

[33] De Bock M, Vandenbroucke RE, Decrock E, Culot M, Cecchelli R, Leybaert L. A new angle on blood-CNS interfaces: A role for connexins? FEBS Letters. 2014;**588**(8):1259-1270

[34] De Bock M, Decrock E, Wang N, Bol M, Vinken M, Bultynck G, et al. The dual face of connexin-based astroglial Ca^{2+} communication: A key player in brain physiology and a prime target in pathology. Biochimica et Biophysica Acta. 2014;**1843**(10):2211-2232

[35] Sáez JC, Berthoud VM, Branes MC, Martinez AD, Beyer EC. Plasma membrane channels formed by connexins: Their regulation and functions. Physiological Reviews. 2003;**83**(4):1359-1400. Epub 2003/09/25

[36] Paul DL, Ebihara L, Takemoto LJ, Swenson KI, Goodenough DA. Connexin46, a novel lens gap junction protein, induces voltage-gated currents in nonjunctional plasma membrane of *Xenopus* oocytes. The Journal of Cell Biology. 1991;**115**(4):1077-1089

[37] Saez JC, Leybaert L. Hunting for connexin hemichannels. FEBS Letters. 2014;**588**(8):1205-1211. Epub 2014/03/19

[38] Orellana JA, von Bernhardi R, Giaume C, Saez JC. Glial hemichannels and their involvement in aging and neurodegenerative diseases. Reviews in the Neurosciences. 2012;**23**(2):163-177. Epub 2012/04/14

[39] Orellana JA, Retamal MA, Moraga-Amaro R, Stehberg J. Role of astroglial hemichannels and pannexons in memory and neurodegenerative diseases. Frontiers in Integrative Neuroscience. 2016;**10**:26. Epub 2016/08/05

[40] Huckstepp RT, Eason R, Sachdev A, Dale N. CO_2-dependent opening of connexin 26 and related beta connexins. The Journal of Physiology. 2010;**588**(Pt 20):3921-3931

[41] Roux L, Madar A, Lacroix MM, Yi C, Benchenane K, Giaume C. Astroglial connexin 43 hemichannels modulate olfactory bulb slow oscillations. The Journal of Neuroscience. 2015;**35**(46):15339-15352. Epub 2015/11/21

[42] Stehberg J, Moraga-Amaro R, Salazar C, Becerra A, Echeverria C, Orellana JA, et al. Release of gliotransmitters through astroglial connexin 43 hemichannels is necessary for fear memory consolidation in the basolateral amygdala. The FASEB Journal. 2012;**26**(9):3649-3657. Epub 2012/06/06

[43] Torres A, Wang F, Xu Q, Fujita T, Dobrowolski R, Willecke K, et al. Extracellular Ca^{2+} acts as a mediator of communication from neurons to glia. Science Signaling. 2012;**5**(208):ra8

[44] Walrave L, Vinken M, Albertini G, De Bundel D, Leybaert L, Smolders IJ. Inhibition of connexin43 hemichannels impairs spatial short-term memory without affecting spatial working memory. Frontiers in Cellular Neuroscience. 2016;**10**:288. Epub 2017/01/10

[45] Dermietzel R, Hertberg EL, Kessler JA, Spray DC. Gap junctions between cultured astrocytes: Immunocytochemical, molecular, and electrophysiological analysis. The Journal of Neuroscience. 1991;**11**(5):1421-1432

[46] Giaume C, Koulakoff A, Roux L, Holcman D, Rouach N. Astroglial networks: A step further in neuroglial and gliovascular interactions. Nature Reviews. Neuroscience. 2010;**11**(2):87-99. Epub 2010/01/21

[47] Kunzelmann P, Schroder W, Traub O, Steinhauser C, Dermietzel R, Willecke K. Late onset and increasing expression of the gap junction protein connexin30 in adult murine brain and long-term cultured astrocytes. Glia. 1999;**25**(2):111-119

[48] Nagy JI, Patel D, Ochalski PA, Stelmack GL. Connexin30 in rodent, cat and human brain: Selective expression in gray matter astrocytes, co-localization with connexin43 at gap junctions and late developmental appearance. Neuroscience. 1999;**88**(2):447-468. Epub 1999/04/10

[49] Yamamoto T, Ochalski A, Hertzberg EL, Nagy JI. LM and EM immunolocalization of the gap junctional protein connexin 43 in rat brain. Brain Research. 1990;**508**(2):313-319. Epub 1990/02/05

[50] Orellana JA, Stehberg J. Hemichannels: New roles in astroglial function. Frontiers in Physiology. 2014;**5**:193. Epub 2014/07/06

[51] Dobrowolski R, Sasse P, Schrickel JW, Watkins M, Kim JS, Rackauskas M, et al. The conditional connexin43G138R mouse mutant represents a new model of hereditary oculodentodigital dysplasia in humans. Human Molecular Genetics. 2008;**17**(4):539-554

[52] Gordon GR, Baimoukhametova DV, Hewitt SA, Rajapaksha WR, Fisher TE, Bains JS. Norepinephrine triggers release of glial ATP to increase postsynaptic efficacy. Nature Neuroscience. 2005;**8**(8):1078-1086. Epub 2005/07/05

[53] Benarroch EE. Adenosine and its receptors: Multiple modulatory functions and potential therapeutic targets for neurologic disease. Neurology. 2008;**70**(3):231-236. Epub 2008/01/16

[54] Morairty S, Rainnie D, McCarley R, Greene R. Disinhibition of ventrolateral preoptic area sleep-active neurons by adenosine: A new mechanism for sleep promotion. Neuroscience. 2004;**123**(2):451-457. Epub 2003/12/31

[55] DeNardo LA, Berns DS, DeLoach K, Luo L. Connectivity of mouse somatosensory and prefrontal cortex examined with trans-synaptic tracing. Nature Neuroscience. 2015;**18**(11):1687-1697

[56] Wang N, De Vuyst E, Ponsaerts R, Boengler K, Palacios-Prado N, Wauman J, et al. Selective inhibition of Cx43 hemichannels by Gap19 and its impact on myocardial ischemia/reperfusion injury. Basic Research in Cardiology. 2013;**108**(1):309. Epub 2012/11/28

Chapter 2

Astrocytic S100B, Blood-Brain Barrier and Neurodegenerative Diseases

Anuradha Krishnan, Hao Wu and Venkat Venkataraman

Abstract

Increased life span and expectations of a better quality of life have resulted in a spotlight on neurodegenerative and cardiovascular diseases generally associated with aging. The drive toward evidence-based medicine has necessitated a constant search for objective biomarkers to assay disease onset, progress, and outcomes to make the best clinical decisions. Enhancement of their use depends on the mechanistic understanding of the biomarker's role in the disease process itself. This chapter focuses on S100B. It is a calcium sensor protein that is primarily astrocytic. While it plays a complex, interlinked role in signaling, serum levels of S100B as a biomarker for clinical decisions is also an area of intense investigation. Both aspects are presented, with an emphasis on the role of S100B in in maintaining a blood-brain barrier, especially in the context of suggesting a unified mechanism for the onset and progression of neurodegenerative diseases.

Keywords: S100B, calcium, blood-brain barrier, biomarker, neurodegeneration, tight junctions

1. Introduction

Rudolph Virchow first proposed the concept of neuroglia as a component of the connective tissue of the brain "nervekitt" [1]. The term "astrocyte" is attributed to Michael von Lenhosseck, coined to denote the stellate (star-like) morphology, with independent contributions also by Kolliker and Anderiezen (reviewed in [2]). The diversity of this group of cells was brought into clear focus by the excellent drawings by Cajal [3]. Glial cells, including astrocytes, were once believed to be limited to passive support in the functioning of the brain. Work over the last few decades has ushered in the understanding that they actively participate in normal metabolism and physiology of the brain, even more so during injury response and repair. They alter the microenvironment through secretion of a variety of signals including cytokines as a result of intracellular process collectively termed "activation," which operates at both ends of time scale—acute and short-term (trauma) as well as chronic and long-term (neurodegenerative diseases). While meant to be adaptive and reparative, they could also lead to exacerbation of injury or disease (for some reviews, please see [4–12]). Understanding the process of activation and its effect on the microenvironment is fundamental to devising positive interventions. One of the important signaling molecules involved in this process is S100B, a calcium sensor protein, which is secreted to act at the extracellular level but also functions intracellularly (reviewed in [13–15]). While primarily astrocytic, it is expressed in other glial cells, non-neuronal cells, and is also detected in the serum. In this chapter, the role for S100B in the neurovascular unit (NVU)—important for the blood-brain barrier (BBB)—is discussed. A summary of conditions in which the serum S100B levels are proposed to be of values as a biomarker provides the backdrop. Based on the findings, a mechanism that places the NVU, with a central role of S100B, at the heart of neurodegenerative diseases is suggested.

2. S100B: the protein

S100B, an astrocytic protein, was originally obtained from the bovine brain [16], as a mixture with S100A1—the fraction was termed "S100" due to partial solubility in 100% saturated ammonium sulfate solution [16], reviewed in [15]. Over 50 years after the identification of S100B, the S100 family now includes more than 20 genes paralogous to S100B, with functions in healthy as well as diseased states [17]. The S100 proteins exist mostly as dimers—homodimers or heterodimers—and share common structural motifs such as the Ca-binding EF hand.

S100B is a homodimer of a 92-amino acid protein, termed S100b, with a molecular mass of 10,713 Da, but migrates between 9 and 14 kDa on SDS-polyacrylamide gels. Crystal structures and refined NMR structures [18, 19] reveal that the monomer contains two EF-hands—one non-canonical and one conventional—that bind calcium; the dimer is in antiparallel orientation. Upon binding calcium, the molecule switches to a more open conformation with hydrophobic patches exposed to facilitate interaction with other molecules. However, these conformational changes are unlike other more conventional EF-containing proteins and are unique for S100B protein—experimentally supported through calcium-induced mobility shift assays [20]. Thus, the S100B protein is specialized to sense changes in calcium levels and mediate appropriate responses, especially in the nervous system.

Expression of S100B in the nervous system is primarily in the glial cells—astrocytes in the central nervous system and Schwann cells in the peripheral nervous system, where they carry out intracellular as well as extracellular functions [reviewed in [14, 15]]. Since these functions are relevant to both healthy and diseased states and S100B is detectable in serum, specifically humans, a focused effort has been underway to determine if S100B could serve as a biomarker.

3. S100B: the biomarker

Biological marker (biomarker) is defined as "a characteristic that is objectively measured and evaluated as an indicator of normal biologic processes, pathological processes or pharmacological responses to a therapeutic intervention" (Biomarker Definitions Working Group). An optimal biomarker is ideally measurable easily, quantitatively related to the extent of the condition, yields reproducible results, and provides guidance regarding outcomes/clinical decisions [21]. The quest for a biomarker may pass through several avenues that include imaging, functional imaging, microRNAs, and microarrays. Even as these journeys have increased the knowledge base, they have also highlighted the complexity of the process. Typically, however, biomarker levels in body fluids—blood, cerebrospinal fluid, saliva, or urine—are more common.

A summary of current information on the correlation of serum S100B levels to different conditions is provided in **Table 1**.

Condition		Citation	Comments
Brain-related			
Traumatic brain injury			
	Mild	[22]	
	Moderate	[23–26]	Not a reliable predictor of good vs. bad outcome
	Severe	[24]	Useful to predict if the patient will regain consciousness 3–6 min after injury, in addition to predicting the outcome
	Children	[27, 28]	Normal serum levels are higher than in adults. Cannot be used as a predictor of injury, but is a predictor of outcome
	Sub-concussive head impacts	[29]	
Non-traumatic intracerebral hemorrhage		[30, 31]	
Stroke (ischemic and hemorrhagic)		[32, 33]	
Heart-related			
Cardiopulmonary bypass surgery		[34]	Meta-analyses
Congestive heart failure		[35]	Further increase if accompanied by renal insufficiency
Dilated cardiomyopathy		[36]	
Skeleton-related			
Orthopedic trauma		[37]	
Hip arthroplasty		[38]	
Cancer-related			
Brain metastases of lung cancer		[39]	
Estrogen receptor-positive breast cancer		[40]	Elevated levels indicate poor disease-free survival
Malignant melanoma		[41–44]	Elevated levels indicate poor disease outcome
Neonates-related			
Congenital heart disease		[45]	Levels dropped to normal 7 days post-operative
Intraventricular hemorrhage, intrauterine growth restrictions, perinatal asphyxia		[21]	
Hypoxic ischemic encephalopathy		[46]	
Other diseases			
Epilepsy		[47, 48]	
Delirium		[49, 50]	
Neuromyelitis optica (AQP4+ve)		[51]	

Condition	Citation	Comments
Neurosarcodiosis	[52]	
Obstructive sleep apnea	[53]	
Proliferative diabetic retinopathy	[54]	
Schizophrenia	[55, 56]	
Systemic lupus erythematous	[57]	

Table 1.
Conditions with elevated serum S100B levels.

In all of the conditions above, serum S100B levels are elevated. It is noted that there are some sporadic reports of decreased levels, particularly in the case of diabetes and anorexia nervosa [58, 59].

The diverse nature of the pathologies with which elevated serum S100B levels are associated, coupled with the lack of functional correlation, has made mechanistic explanations difficult to say the least. While S100B is currently known to be expressed in multiple tissues, the primary source of serum S100B is believed to be of astrocytic origin [60, 61]. Therefore, for S100B levels in serum to be elevated, astrocytic S100B must be able to reach the blood, which would not normally happen due to the presence of the blood-brain barrier (BBB).

The BBB comprises the physiological and functional barrier that separates the nervous system from the circulatory system [62, 63] and its breakdown allows for the leakage of damaging humoral elements into the brain parenchyma [63–67].

4. S100B and blood-brain barrier (BBB)

4.1 BBB and neurodegenerative diseases

The BBB is mainly formed by a monolayer of brain vascular endothelial cells (BVECs) that are sealed by tight junctions (TJs) [reviewed in [68–70]]. It actively regulates transportation of metabolic wastes and nutrients, such as ions, glucose, and amino acids, between blood and brain interstitial fluid (ISF) [reviewed in [69]]. On the other hand, other plasma components such as immunoglobulins and cells such as leukocytes are restricted. Thus, the BBB enables neurons and supporting cells to receive nutrients and remove wastes. In addition, they provide protection from the immune system. The concept of the neurovascular unit (NVU), to maintain the function of the BBB in health and underlie its response during disease, has been proposed [71–74].

Impaired function of the BBB has been linked to a variety of pathological conditions that affect the brain [reviewed in [62, 63, 67, 75]]. These include epilepsy [reviewed in [76]], psychiatric diseases such as neuropsychiatric lupus [77], dementias such as Parkinson's disease [78–80], Alzheimer's disease (AD) [81–86], other neurodegenerative diseases such as Huntington's disease [87], amyotrophic lateral sclerosis [88, 89], multiple sclerosis [90, 91], and those caused by viral infections [92–94]. The compromise of the BBB has been tightly linked causally or as a diagnostic marker to chronic neurodegenerative diseases such as AD [95–97] and acute conditions such as delirium [98]. Yet, the causes and consequences of the BBB breach in those diseases remain elusive. Work carried out in this laboratory has established a role for S100B in maintaining an intact BBB using a mouse (S100BKO) model [66].

4.2 S100B is essential to maintain BBB

The detection of leaked serum components, such as IgG, has been widely used to assess impairment of BBB function [64]. In a study from this laboratory by Wu and coworkers [66], vascular leaks in the brain were evaluated by immunostaining brain sections from S100BKO mice to detect extravascular IgG. Previous studies have shown that intravenous injection of pertussis toxin [PT] generates leaks in the BBB [64, 99]. Therefore, wild type and S100BKO mice were injected with PT and the effect on BBB permeability was investigated.

Extravasated IgG from the blood vessels was detected as perivascular leakage clouds marking sites of BBB breach. In wild-type mice, leak clouds were detected only upon injection of PT; in the S100BKO mice, however, they were detected even without PT injection by 6 months of age and were exacerbated by PT injection. Thus, there is an endogenous BBB deficiency in S100BKO mice, which is increased by treatment with PT. The BBB breach was chronic and age-dependent, increasing with age [66]. Thus, the system mimics the chronic, age-associated compromise of BBB in humans. In addition, selective binding of neurons by IgG was also temporally and spatially associated with these leak clouds, suggesting that neuron-binding autoantibodies are present and BBB compromise allows their access to neurons in the brain; an increase in the brain-reactive autoantibodies was also associated with increased BBB breach [66].

Interestingly, despite detectable pathology, there was very little glial response as measured by increased expression of glial fibrillary acidic protein (GFAP) [66]. Is it possible that S100B is necessary to trigger the astroglial response to neuronal injury/insult? The question remains to be answered.

The potential reason for the BBB breach, however, could be identified: Based on electron microscopic analyses, disorganization of endothelial tight junctions is proposed to cause the observed BBB breach [66].

4.3 S100B is essential to maintain blood-retinal barrier (BRB)

Tight junctions are also important for the maintenance of the blood-retinal barrier (BRB). The existence of blood-retinal barrier (BRB) is well established [100–104], although its similarity to the BBB remains to be completely elucidated [105, 106] and its establishment, less understood [107]. If S100B was essential for the maintenance of tight junctions in vascular endothelial cells and lack of it caused BBB breach, would similar breaches be observed BRB also? The results presented below show that this is, indeed, the case.

In untreated wild-type mice (**Figure 1A**), staining for IgG was restricted to within blood vessels, which indicates an intact barrier. In PT-injected wild-type mice, IgG extravasated from the blood vessels. However, the perivascular leakage clouds were not obvious while the IgG-bound ganglion cells were detected (indicated by arrows in **Figure 1B**). The results suggest that neuron-binding autoantibodies are present and BRB compromise allows their access to neurons in the retina. Leak clouds outside the blood vessels were observed in S100BKO mice in the absence of PT (**Figure 1C**). Upon PT injection of S100BKO mice, the perivascular leak clouds persist (**Figure 1D**). Thus, an endogenous BRB deficiency in S100BKO mice was observed. The presence of retina-specific autoantibodies was confirmed independently by Western blot analyses (**Figure 1E**). The result shows that their appearance is age-dependent, as in the BBB.

Thus, the effect of S100B deprivation leads to chronic barrier disruption in both brain and retina through disruption of the endothelial tight junctions, most likely. The details of the mechanistic aspects of the action of S100B on tight junction maintenance remain to be established: both intracellular and extracellular routes are possible [14]. An intracellular mechanism is supported by observations that S100B is

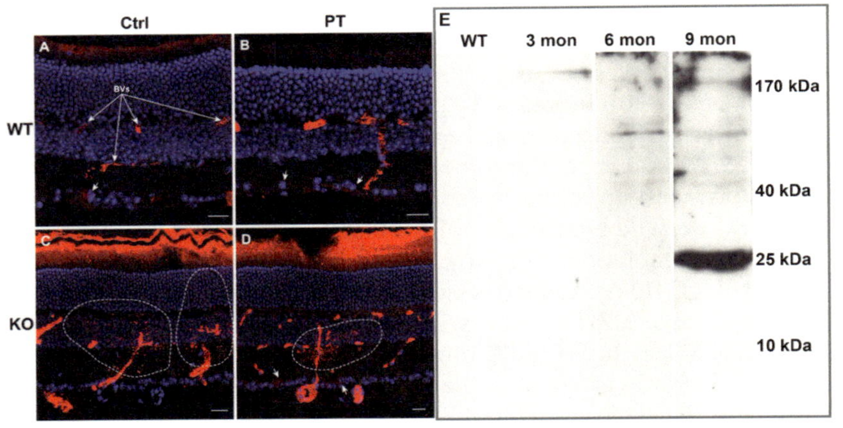

Figure 1.
S100BKO mice demonstrate significant BRB compromise in the retina and express retina-specific autoantibodies. Overlay of IgG immunostaining (red) with DAPI (blue) is presented from the retinal sections of untreated wild-type mice (A), PT-treated WT mice (B), S100BKO mice (C), and S100BKO mice treated with PT (D). Scale bar, 20 μm. Western blots (E) of the swine retinal protein extract were probed with pooled sera from wild-type mice (WT) or from S100BKO mice at 3 (3 Mon), 6 (6 Mon), or 9 (9 Mon) months of age. A representative result is shown. Molecular size markers are indicated alongside.

expressed in endothelial cells [108, 109]; furthermore, assembling and maintaining functional tight junctions is dependent upon several signaling pathways [reviewed in [110–112]], all of which are known to be influenced by S100B: calcium homeostasis [13, 14, 113], guanylate cyclase activation [114, 115], modulation of rhoGTPases such as Rac1 [116] and protein kinase C activity [108]. Extracellularly, S100B, most likely, acts through the receptor for advanced glycation end products (RAGE) expressed on endothelial cells [117] and results in the activation of the Ras-ERK1/2-NF-κB pathway [reviewed in [14]]. This pathway regulates endothelial hyperpermeability [reviewed in [111]], particularly the assembly of TJ proteins [118]. Moreover, the expression of NF-κB itself is regulated by S100B [119, 120]. Additional receptors for S100B, such as CD166/ALCAM and Toll-Like Receptors (TLR), are also known. Therefore, extracellular S100B may also be critical for endothelial function in maintaining the BBB.

5. Blood-brain barrier (BBB) and neurodegenerative diseases

5.1 Autoantibodies and neurodegenerative diseases

The BBB breach allows autoantibodies in the blood vessels to gain access to neurons and other cell types, causing compromise in function and, in extreme cases, cell death. The breach itself could be generated over time through age or disorders or acutely through traumatic injuries. Debris released from sick or dead cells would now be encountered by the immune system, which would mount an antibody response, generating autoantibodies. This results in a potentially devastating feedback loop: the autoantibodies cause compromise of cells, releasing debris, which, in turn, augments the response. The only way to halt this cycle will be through the loss of the targeted antigens: either through endocytosis or receptor stripping. Typically, that would also lead to dysfunctions in neurons and other cells [121, 122].

This "positive feedback loop" hypothesis is suggested by a decreased expression of MAP2, shown to be indicative of neuronal stress [123, 124], in both *in vivo* [66] and *in vitro* model systems of BBB [125] or BRB [126] breach. Previous studies have also shown a widespread presence of brain-reactive autoantibodies in human serum [127] and changing autoantibody profiles upon disease [96, 122, 128–130]. Much evidence now suggests that BBB breakdown also contributes to acute dysfunctions such as post-operative delirium and recovery from anesthesia [131, 132]. An increase in BBB permeability, like in S100BKO mice reported here, has been observed previously in humans with age [82] and with many chronic dysfunctions **Table 1**. Therefore, the S100BKO mouse may serve as a useful model to mimic the status of the aged BBB. It is well documented that the elderly population is highly susceptible to neurodegenerative diseases and delirium.

Increased levels of autoimmune antibodies (generally associated with increased BBB permeability) have been reported in several diseases (reviewed in [129–134]). A positive correlation between autoantibody prevalence and age/diseased state has propelled the idea that they are potential diagnostic tools in AD and Parkinson's disease [122, 129–136].

5.2 Toward a unified mechanism

A clear understanding of the structural components and functions of the BBB may be the key to delineating pathologies of the brain. Here, we propose the idea that neurodegeneration is a multi-step process (**Figure 2**) involving BBB breach,

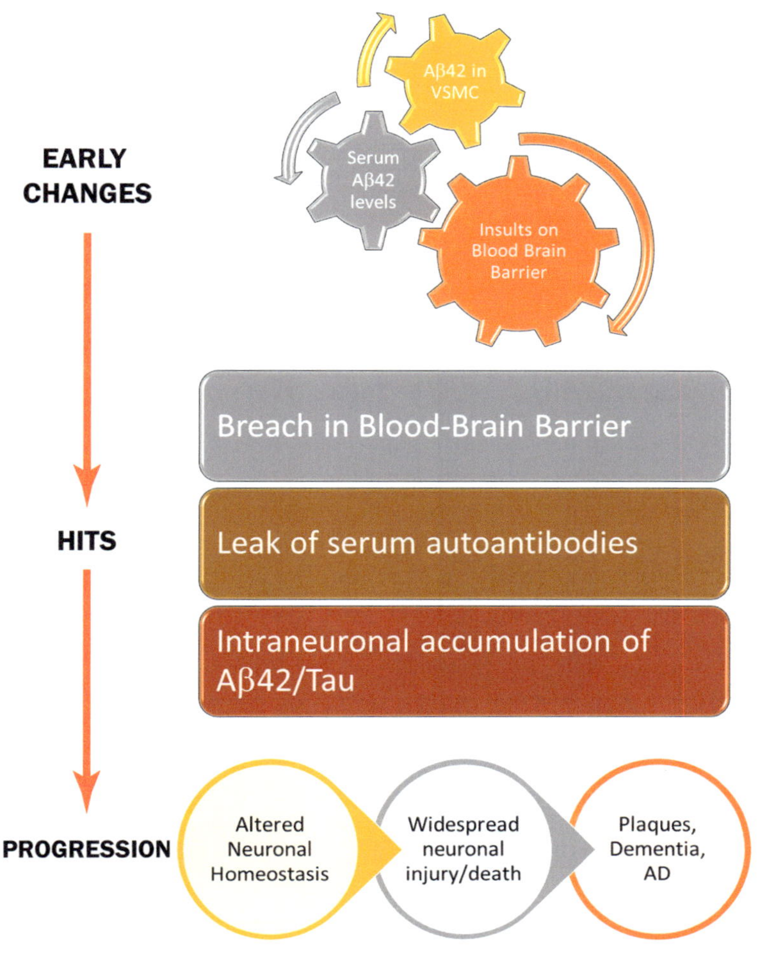

Figure 2.
Blood-brain barrier breach, autoantibodies and neurodegeneration. A unified mechanism is proposed for neurodegenerative diseases. Early changes include changes in the NVU, serum, and other insults on the BBB. Once the BBB is breached, it leads to the extravasation of serum components and access/production of brain-reactive auto antibodies. This results in a positive feedback loop, altering homeostasis and eventually resulting in the disease phenotype through neuronal injury/death.

infiltration of auto antibodies and other damaging plasma components into the brain parenchyma with death of the neurons as the long-term sequelae. Multiple studies indicate that the BBB is very important to the brain health, neuronal integrity, and homeostasis; when BBB breach occurs, it allows for the extravasation of blood-borne molecules (such as Aβ42 in AD), brain-reactive antibodies, and inflammatory factors into the normally immune-privileged brain parenchyma [127, 137–139]. Access of the previously excluded and potentially damaging blood-borne plasma elements to the brain interstitium results in disruption of brain homeostasis, impaired neuronal function, and eventually, neuronal loss [64, 134, 140, 141]. Furthermore, injury or disease of the central nervous system (CNS), such as AD, causes gliosis, which is characterized by the activation of astrocytes, microglia, and other cell types.

Insults to the BBB can be brought about by several pathological conditions and result in the compromise of this protective layer. Studies from our lab show that the inner blood-retinal barrier (BRB) is very similar to the BBB and can be used as a model system to study BBB [126]. The use of brain-reactive autoantibodies to diagnose neurodegenerative diseases with a high degree of confidence has also been reported [122, 135, 136, 142–145].

Taken together, investigations into the BBB maintenance will yield rich dividends toward the mechanistic understanding that may underlie multiple neurodegenerative diseases, increased diagnostic tools in terms of model systems and biomarkers and, perhaps, also drug delivery options. Delineation of S100B signaling pathways is likely to contribute significantly toward this end.

6. Conclusions

S100B, primarily astrocytic in origin, is a unique signaling molecule that impacts multiple signaling pathways—sometimes negatively, sometimes positively [15, 146]. The dual nature of action—intra- and extracellular—poses a significant challenge in delineating the precise mechanism of action in many instances. Yet, S100B is emerging as a central molecule (**Figure 3**) in regulating normal and

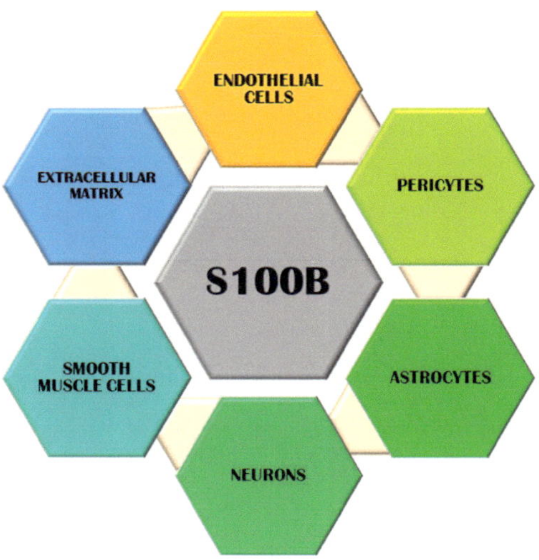

Figure 3.
A central role for S100B, an astrocytic protein. The figure depicts the multiple cell-types contributing to an intact BBB that form the NVU. By virtue of being an extracellular signal as well as being expressed intracellularly, S100B is proposed to play a central role in maintaining BBB and serve as a marker for neurodegeneration.

disease processes—especially where astrocytes/glial cells are involved—the enteric glial cells being a recent and exciting example [147]. It is hoped that a fundamental mechanistic understanding would enable decipher the process, refine the clinical relevance biomarker, and carry the laboratory bench knowledge to the patient bedside to improve quality of life and clinical interventions.

Acknowledgements

The authors are extremely grateful to excellent discussions with Professor Robert Nagele. The support for research through grants from Osteopathic Heritage Foundation and American Osteopathic Association is gratefully acknowledged.

Author details

Anuradha Krishnan[1†], Hao Wu[2†] and Venkat Venkataraman[1,2,3*]

1 Department of Cell Biology and Neuroscience, Rowan School of Osteopathic Medicine, Rowan University, Stratford, New Jersey, USA

2 Graduate School of Biomedical Sciences, Rowan School of Osteopathic Medicine, Rowan University, Stratford, New Jersey, USA

3 Department of Rehabilitation Medicine, Neuro-Musculoskeletal Institute, Rowan School of Osteopathic Medicine, Rowan University, Stratford, New Jersey, USA

*Address all correspondence to: venkatar@rowan.edu

† These authors have contributed equally.

References

[1] Virchow R. Über das granulierte Ansehen der Wandlungen der Hirnventrikel. Allgemeine Zeitschrift für Psychiatrie und Psychisch-Gerichtliche Medizin. 1846;**3**:242

[2] Kettenmann H, Verkhratsky A. Neuroglia: The 150 years after. Trends in Neurosciences. 2008;**31**(12):653-659

[3] Cajal SR. Histology of the nervous system of man and vertebrates. In: History of Neuroscience. New York: Oxford University Press; 1995 (6)

[4] Allen NJ. Astrocyte regulation of synaptic behavior. Annual Review of Cell and Developmental Biology. 2014;**30**:439-463

[5] Butt A, Verkhratsky A. Neuroglia: Realising their true potential. Brain and Neuroscience Advances. 2018;**2**:1-6. DOI: 10.1177/2398212818817495

[6] Greenhalgh AD, David S, Bennett FC. Immune cell regulation of glia during CNS injury and disease. Nature Reviews. Neuroscience. 2020;**21**(3):139-152

[7] McConnell HL, Kersch CN, Woltjer RL, Neuwelt EA. The translational significance of the neurovascular unit. The Journal of Biological Chemistry. 2017;**292**(3): 762-770

[8] Michinaga S, Koyama Y. Dual roles of astrocyte-derived factors in regulation of blood-brain barrier function after brain damage. International Journal of Molecular Sciences. 2019;**20**(3):571-593

[9] Ponath G, Park C, Pitt D. The role of astrocytes in multiple sclerosis. Frontiers in Immunology. 2018;**9**:217

[10] Spampinato SF, Bortolotto V, Canonico PL, Sortino MA, Grilli M. Astrocyte-derived paracrine signals: Relevance for neurogenic niche regulation and blood-brain barrier integrity. Frontiers in Pharmacology. 2019;**10**:1346

[11] Takahashi S. Metabolic compartmentalization between astroglia and neurons in physiological and pathophysiological conditions of the neurovascular unit. Neuropathology. 2020;**40**:121-137

[12] Vainchtein ID, Molofsky AV. Astrocytes and microglia: In sickness and in health. Trends in Neurosciences. 2020;**43**(3):144-154

[13] Donato R, Cannon BR, Sorci G, Riuzzi F, Hsu K, Weber DJ, et al. Functions of S100 proteins. Current Molecular Medicine. 2013;**13**(1):24-57

[14] Donato R, Sorci G, Riuzzi F, Arcuri C, Bianchi R, Brozzi F, et al. S100B's double life: Intracellular regulator and extracellular signal. Biochimica et Biophysica Acta. 2009;**1793**(6):1008-1022

[15] Van Eldik LJ, Wainwright MS. The Janus face of glial-derived S100B: Beneficial and detrimental functions in the brain. Restorative Neurology and Neuroscience. 2003;**21**(3-4):97-108

[16] Moore BW. A soluble protein characteristic of the nervous system. Biochemical and Biophysical Research Communications. 1965;**19**(6):739-744

[17] Gonzalez LL, Garrie K, Turner MD. Role of S100 proteins in health and disease. Biochimica et Biophysica Acta (BBA): Molecular Cell Research. 1867;**2020**(6):118677

[18] Matsumura H, Shiba T, Inoue T, Harada S, Kai Y. A novel mode of target recognition suggested by the 2.0 A structure of holo S100B from bovine brain. Structure. 1998;**6**(2):233-241

[19] Wright NT, Inman KG, Levine JA, Cannon BR, Varney KM, Weber DJ. Refinement of the solution structure and dynamic properties of Ca^{2+}-bound rat S100B. Journal of Biomolecular NMR. 2008;**42**(4):279-286

[20] Viviano J, Krishnan A, Wu H, Venkataraman V. Electrophoretic mobility shift in native gels indicates calcium-dependent structural changes of neuronal calcium sensor proteins. Analytical Biochemistry. 2016;**494**:93-100

[21] Bersani I, Pluchinotta F, Dotta A, Savarese I, Campi F, Auriti C, et al. Early predictors of perinatal brain damage: The role of neurobiomarkers. Clinical Chemistry and Laboratory Medicine. 2020;**58**(4):471-486

[22] Jones CMC, Harmon C, McCann M, Gunyan H, Bazarian JJ. S100B outperforms clinical decision rules for the identification of intracranial injury on head CT scan after mild traumatic brain injury. Brain Injury. 2020:1-8

[23] Dadas A, Washington J, Diaz-Arrastia R, Janigro D. Biomarkers in traumatic brain injury (TBI): A review. Neuropsychiatric Disease and Treatment. 2018;**14**:2989-3000

[24] Kawata K, Liu CY, Merkel SF, Ramirez SH, Tierney RT, Langford D. Blood biomarkers for brain injury: What are we measuring? Neuroscience & Biobehavioral Reviews. 2016;**68**:460-473

[25] Abbasi M, Sajjadi M, Fathi M, Maghsoudi M. Serum S100B protein as an outcome prediction tool in emergency department patients with traumatic brain injury. Turkish Journal of Emergency Medicine. 2014;**14**(4):147-152

[26] Wang KK, Yang Z, Zhu T, Shi Y, Rubenstein R, Tyndall JA, et al. An update on diagnostic and prognostic biomarkers for traumatic brain injury. Expert Review of Molecular Diagnostics. 2018;**18**(2):165-180

[27] Kovesdi E, Luckl J, Bukovics P, Farkas O, Pal J, Czeiter E, et al. Update on protein biomarkers in traumatic brain injury with emphasis on clinical use in adults and pediatrics. Acta Neurochirurgica. 2010;**152**(1):1-17

[28] Manzano S, Holzinger IB, Kellenberger CJ, Lacroix L, Klima-Lange D, Hersberger M, et al. Diagnostic performance of S100B protein serum measurement in detecting intracranial injury in children with mild head trauma. Emergency Medicine Journal. 2016;**33**(1):42-46

[29] Zonner SW, Ejima K, Bevilacqua ZW, Huibregtse ME, Charleston C, Fulgar C, et al. Association of increased serum S100B levels with high school football subconcussive head impacts. Frontiers in Neurology. 2019;**10**:327

[30] Aydin I, Algin A, Poyraz MK, Yumrutas O. Diagnostic value of serum glial fibrillary acidic protein and S100B serum levels in emergency medicine patients with traumatic versus nontraumatic intracerebral hemorrhage. Nigerian Journal of Clinical Practice. 2018;**21**(12):1645-1650

[31] Ferrete-Araujo AM, Rodriguez-Rodriguez A, Egea-Guerrero JJ, Vilches-Arenas A, Godoy DA, Murillo-Cabezas F. Brain injury biomarker behavior in spontaneous intracerebral hemorrhage. World Neurosurgery. 2019;**132**:e496-e505

[32] Montaner J, Mendioroz M, Delgado P, García-Berrocoso T, Giralt D, Merino C, et al. Differentiating ischemic from hemorrhagic stroke using plasma biomarkers: The S100B/RAGE pathway. Journal of Proteomics. 2012;**75**(15):4758-4765

[33] Glushakova OY, Glushakov AV, Miller ER, Valadka AB, Hayes RL.

Biomarkers for acute diagnosis and management of stroke in neurointensive care units. Brain Circulation. 2016;**2**(1):28-47

[34] Yuan SM. Biomarkers of cerebral injury in cardiac surgery. Anadolu Kardiyoloji Dergisi. 2014;**14**(7):638-645

[35] Li JP, Lu L, Wang LJ, Zhang FR, Shen WF. Increased serum levels of S100B are related to the severity of cardiac dysfunction, renal insufficiency and major cardiac events in patients with chronic heart failure. Clinical Biochemistry. 2011;**44**(12):984-988

[36] Mazzini GS, Schaf DV, Vinade ER, Horowitz E, Bruch RS, Brunm LM, et al. Increased S100B serum levels in dilated cardiomyopathy patients. Journal of Cardiac Failure. 2007;**13**(10):850-854

[37] Papa L, Silvestri S, Brophy GM, Giordano P, Falk JL, Braga CF, et al. GFAP out-performs S100beta in detecting traumatic intracranial lesions on computed tomography in trauma patients with mild traumatic brain injury and those with extracranial lesions. Journal of Neurotrauma. 2014;**31**(22):1815-1822

[38] Tomaszewski D, Balkota M, Rybicki Z. Regional cerebral oxygen saturation decreases during primary hip arthroplasty: An analysis of perioperative regional cerebral oxygenation (rSO_2), S100 calcium-binding protein B (S100B) and glial fibrillary acidic protein (GFAP) values. A pilot study. Medical Science Monitor. 2019;**25**:525-531

[39] Choi H, Puvenna V, Brennan C, Mahmoud S, Wang XF, Phillips M, et al. S100B and S100B autoantibody as biomarkers for early detection of brain metastases in lung cancer. Translational Lung Cancer Research. 2016;**5**(4):413-419

[40] Charmsaz S, Hughes E, Bane FT, Tibbitts P, McIlroy M, Byrne C, et al. S100beta as a serum marker in endocrine resistant breast cancer. BMC Medicine. 2017;**15**(1):79

[41] Harpio R, Einarsson R. S100 proteins as cancer biomarkers with focus on S100B in malignant melanoma. Clinical Biochemistry. 2004;**37**(7):512-518

[42] Martenson ED, Hansson LO, Nilsson B, von Schoultz E, Mansson Brahme E, Ringborg U, et al. Serum S-100b protein as a prognostic marker in malignant cutaneous melanoma. Journal of Clinical Oncology. 2001;**19**(3):824-831

[43] Karonidis A, Mantzourani M, Gogas H, Tsoutsos D. Serum S100B levels correlate with stage, N status, mitotic rate and disease outcome in melanoma patients independent to LDH. Journal of BUON. 2017;**22**(5):1296-1302

[44] Ertekin SS, Podlipnik S, Ribero S, Molina R, Rios J, Carrera C, et al. Monthly changes in serum levels of S100B protein as a predictor of metastasis development in high-risk melanoma patients. Journal of the European Academy of Dermatology and Venereology. 2020. In press. DOI: 10.1111/jdv.16212

[45] Trakas E, Domnina Y, Panigrahy A, Baust T, Callahan PM, Morell VO, et al. Serum neuronal biomarkers in neonates with congenital heart disease undergoing cardiac surgery. Pediatric Neurology. 2017;**72**:56-61

[46] Zaigham M, Lundberg F, Olofsson P. Protein S100B in umbilical cord blood as a potential biomarker of hypoxic-ischemic encephalopathy in asphyxiated newborns. Early Human Development. 2017;**112**:48-53

[47] Liang KG, Mu RZ, Liu Y, Jiang D, Jia TT, Huang YJ. Increased serum S100B levels in patients with epilepsy:

A systematic review and meta-analysis study. Frontiers in Neuroscience. 2019;**13**:456

[48] Meguid NA, Samir H, Bjorklund G, Anwar M, Hashish A, Koura F, et al. Altered S100 calcium-binding protein B and matrix metallopeptidase 9 as biomarkers of mesial temporal lobe epilepsy with hippocampus sclerosis. Journal of Molecular Neuroscience. 2018;**66**(4):482-491

[49] van Munster BC, Korevaar JC, Korse CM, Bonfrer JM, Zwinderman AH, de Rooij SE. Serum S100B in elderly patients with and without delirium. International Journal of Geriatric Psychiatry. 2010;**25**(3):234-239

[50] Li Y, Yu ZX, Ji MS, Yan J, Cai Y, Liu J, et al. A pilot study of the use of dexmedetomidine for the control of delirium by reducing the serum concentrations of brain-derived neurotrophic factor, neuron-specific enolase, and S100B in polytrauma patients. Journal of Intensive Care Medicine. 2019;**34**(8):674-681

[51] Fujii C, Tokuda T, Ishigami N, Mizuno T, Nakagawa M. Usefulness of serum S100B as a marker for the acute phase of aquaporin-4 autoimmune syndrome. Neuroscience Letters. 2011;**494**(1):86-88

[52] Moss BP, Patel DC, Tavee JO, Culver DA. Evaluating S100B as a serum biomarker for central neurosarcoidosis. Respiratory Medicine. 2020;**162**:105855

[53] Rezaei F, Abbasi H, Sadeghi M, Imani MM. The effect of obstructive sleep apnea syndrome on serum S100B and NSE levels: A systematic review and meta-analysis of observational studies. BMC Pulmonary Medicine. 2020;**20**(1):31

[54] Asadova V, Gul Z, Buyukuysal RL, Yalcinbayir O. Assessment of neuron-specific enolase, S100B and malondialdehyde levels in serum and vitreous of patients with proliferative diabetic retinopathy. International Ophthalmology. 2020;**40**(1):227-234

[55] Sen J, Belli A. S100B in neuropathologic states: The CRP of the brain? Journal of Neuroscience Research. 2007;**85**(7):1373-1380

[56] Yelmo-Cruz S, Morera-Fumero AL, Abreu-Gonzalez P. S100B and schizophrenia. Psychiatry and Clinical Neurosciences. 2013;**67**(2):67-75

[57] Noris-Garcia E, Arce S, Nardin P, Lanigan ME, Acuna V, Gutierrez F, et al. Peripheral levels of brain-derived neurotrophic factor and S100B in neuropsychiatric systemic lupus erythematous. Lupus. 2018;**27**(13):2041-2049

[58] Holtkamp K, Buhren K, Ponath G, von Eiff C, Herpertz-Dahlmann B, Hebebrand J, et al. Serum levels of S100B are decreased in chronic starvation and normalize with weight gain. Journal of Neural Transmission (Vienna). 2008;**115**(6):937-940

[59] Riuzzi F, Chiappalupi S, Arcuri C, Giambanco I, Sorci G, Donato R. S100 proteins in obesity: Liaisons dangereuses. Cellular and Molecular Life Sciences. 2020;**77**(1):129-147

[60] Astrand R, Unden J. Clinical use of the calcium-binding S100B protein, a biomarker for head injury. Methods in Molecular Biology. 1929;**2019**:679-690

[61] Marchi N, Rasmussen P, Kapural M, Fazio V, Kight K, Mayberg MR, et al. Peripheral markers of brain damage and blood-brain barrier dysfunction. Restorative Neurology and Neuroscience. 2003;**21**(3-4):109-121

[62] Sharif Y, Jumah F, Coplan L, Krosser A, Sharif K,

Tubbs RS. Blood brain barrier: A review of its anatomy and physiology in health and disease. Clinical Anatomy. 2018;**31**(6):812-823

[63] Sweeney MD, Zhao Z, Montagne A, Nelson AR, Zlokovic BV. Blood-brain barrier: From physiology to disease and back. Physiological Reviews. 2019;**99**(1):21-78

[64] Clifford PM, Zarrabi S, Siu G, Kinsler KJ, Kosciuk MC, Venkataraman V, et al. Abeta peptides can enter the brain through a defective blood-brain barrier and bind selectively to neurons. Brain Research. 2007;**1142**:223-236

[65] Nagele RG, Acharya NK, Han M, DeMarshall C, Kosciuk MC, Nagele EP. Blood-brain barrier (Bbb) breakdown and brain-reactive autoantibodies as a trigger for Cns diseases. Gerontologist. 2012;**52**:453

[66] Wu H, Brown EV, Acharya NK, Appelt DM, Marks A, Nagele RG, et al. Age-dependent increase of blood-brain barrier permeability and neuron-binding autoantibodies in S100B knockout mice. Brain Research. 1637;**2016**:154-167

[67] Zhang W, Zhu L, An C, Wang R, Yang L, Yu W, et al. Brain hemorrhages: The blood brain barrier in cerebral ischemic injury—Disruption and repair. Brain Hemorrhages. 2020. In Press. DOI: 10.1016/j.hest.2019.12.004

[68] Abbott NJ, Rönnbäck L, Hansson E. Astrocyte–endothelial interactions at the blood–brain barrier. Nature Reviews Neuroscience. 2006;7(1):41-53

[69] Abbott NJ, Patabendige AA, Dolman DE, Yusof SR, Begley DJ. Structure and function of the blood–brain barrier. Neurobiology of Disease. 2010;**37**(1):13-25

[70] Zlokovic BV. Neurovascular pathways to neurodegeneration in Alzheimer's disease and other disorders. Nature Reviews Neuroscience. 2011;**12**(12):723-738

[71] Brown WR, Thore CR. Review: Cerebral microvascular pathology in ageing and neurodegeneration. Neuropathology and Applied Neurobiology. 2011;**37**(1):56-74

[72] Chao YX, He BP, Tay SSW. Mesenchymal stem cell transplantation attenuates blood brain barrier damage and neuroinflammation and protects dopaminergic neurons against MPTP toxicity in the substantia nigra in a model of Parkinson's disease. Journal of Neuroimmunology. 2009;**216**(1):39-50

[73] Hawkins BT, Davis TP. The blood-brain barrier/neurovascular unit in health and disease. Pharmacological Reviews. 2005;**57**(2):173-185

[74] Zlokovic BV. Neurodegeneration and the neurovascular unit. Nature Medicine. 2010;**16**(12):1370-1371

[75] Sweeney MD, Sagare AP, Zlokovic BV. Blood-brain barrier breakdown in Alzheimer disease and other neurodegenerative disorders. Nature Reviews. Neurology. 2018;**14**(3):133-150

[76] Gorter JA, van Vliet EA, Aronica E. Status epilepticus, blood-brain barrier disruption, inflammation, and epileptogenesis. Epilepsy & Behavior. 2015;**46**:13-16

[77] Stock AD, Wen J, Putterman C. Neuropsychiatric lupus, the blood brain barrier, and the TWEAK/Fn14 pathway. Frontiers in Immunology. 2013;**4**:484

[78] Cabezas R, Ávila M, Gonzalez J, El-Bachá RS, Báez E, García-Segura LM, et al. Astrocytic modulation of blood brain barrier: Perspectives on

Parkinson's disease. Frontiers in Cellular Neuroscience. 2014;**8**:211

[79] Kortekaas R, Leenders KL, van Oostrom JC, Vaalburg W, Bart J, Willemsen AT, et al. Blood-brain barrier dysfunction in parkinsonian midbrain in vivo. Annals of Neurology. 2005;**57**(2):176-179

[80] Pisani V, Stefani A, Pierantozzi M, Natoli S, Stanzione P, Franciotta D, et al. Increased blood-cerebrospinal fluid transfer of albumin in advanced Parkinson's disease. Journal of Neuroinflammation. 2012;**9**:188

[81] Clifford PM, Zarrabi S, Siu G, Kinsler KJ, Kosciuk MC, Venkataraman V, et al. Aβ peptides can enter the brain through a defective blood–brain barrier and bind selectively to neurons. Brain Research. 2007;**1142**:223-236

[82] Farrall AJ, Wardlaw JM. Blood–brain barrier: Ageing and microvascular disease—Systematic review and meta-analysis. Neurobiology of Aging. 2009;**30**(3):337-352

[83] Sengillo JD, Winkler EA, Walker CT, Sullivan JS, Johnson M, Zlokovic BV. Deficiency in mural vascular cells coincides with blood-brain barrier disruption in Alzheimer's disease. Brain Pathology (Zurich, Switzerland). 2013;**23**(3):303-310

[84] de la Torre JC. Is Alzheimer's disease a neurodegenerative or a vascular disorder? Data, dogma, and dialectics. The Lancet Neurology. 2004;**3**(3):184-190

[85] De La Torre JC. The vascular hypothesis of Alzheimer's disease: Bench to bedside and beyond. Neurodegenerative Diseases. 2010;**7**(1-3):116-121

[86] de la Torre JC. Cerebral hemodynamics and vascular risk factors: Setting the stage for Alzheimer's disease. Journal of Alzheimer's Disease. 2012;**32**(3):553-567

[87] Drouin-Ouellet J, Sawiak SJ, Cisbani G, Lagace M, Kuan WL, Saint-Pierre M, et al. Cerebrovascular and blood-brain barrier impairments in Huntington's disease: Potential implications for its pathophysiology. Annals of Neurology. 2015;**78**:160-177

[88] Garbuzova-Davis S, Hernandez-Ontiveros DG, Rodrigues MC, Haller E, Frisina-Deyo A, Mirtyl S, et al. Impaired blood–brain/spinal cord barrier in ALS patients. Brain Research. 2012;**1469**:114-128

[89] Winkler EA, Sengillo JD, Sullivan JS, Henkel JS, Appel SH, Zlokovic BV. Blood-spinal cord barrier breakdown and pericyte reductions in amyotrophic lateral sclerosis. Acta Neuropathologica. 2013;**125**(1):111-120

[90] Bennett J, Basivireddy J, Kollar A, Biron KE, Reickmann P, Jefferies WA, et al. Blood–brain barrier disruption and enhanced vascular permeability in the multiple sclerosis model EAE. Journal of Neuroimmunology. 2010;**229**(1):180-191

[91] Larsson H, Stubgaard M, Frederiksen J, Jensen M, Henriksen O, Paulson O. Quantitation of blood-brain barrier defect by magnetic resonance imaging and gadolinium-DTPA in patients with multiple sclerosis and brain tumors. Magnetic Resonance in Medicine. 1990;**16**(1):117-131

[92] Calderón-Peláez M-A, Velandia-Romero ML, Bastidas-Legarda LY, Beltrán EO, Camacho-Ortega SJ, Castellanos JE. Dengue virus infection of blood–brain barrier cells: Consequences of severe

disease. Frontiers in Microbiology. 2019;**10**(1435):1-15

[93] Leda AR, Bertrand L, Andras IE, El-Hage N, Nair M, Toborek M. Selective disruption of the blood–brain barrier by Zika virus. Frontiers in Microbiology. 2019;**10**(2158):1-14

[94] Li F, Wang Y, Yu L, Cao S, Wang K, Yuan J, et al. Viral infection of the central nervous system and neuroinflammation precede blood-brain barrier disruption during Japanese encephalitis virus infection. Journal of Virology. 2015;**89**(10):5602-5614

[95] Di Marco LY, Venneri A, Farkas E, Evans PC, Marzo A, Frangi AF. Vascular dysfunction in the pathogenesis of Alzheimer's disease—A review of endothelium-mediated mechanisms and ensuing vicious circles. Neurobiology of Disease. 2015;**82**:593-606

[96] Nagele RG, Clifford PM, Siu G, Levin EC, Acharya NK, Han M, et al. Brain-reactive autoantibodies prevalent in human sera increase intraneuronal amyloid-beta(1-42) deposition. Journal of Alzheimer's Disease. 2011;**25**(4):605-622

[97] Thomas J, Lewis CLT. The End of Alzheimer's?: A Differential Diagnosis Toward a Cure. TJLPHD, LLC; 2014. pp. 471

[98] Goldwaser E, Acharya N, Nagele R. Cerebrovascular and blood-brain barrier compromise: A mechanistic link between vascular disease and Alzheimer's disease subtypes of neurocognitive disorders. Journal of Parkinson's Disease and Alzheimer's Disease. 2015;**2**(2):10

[99] Munoz JJ. Biological activities of pertussigen (pertussis toxin). Pertussis Toxin. 1985:1-18

[100] Cunha-Vaz J, Bernardes R, Lobo C. Blood-retinal barrier. European Journal of Ophthalmology. 2011;**21**(Suppl 6):S3-S9

[101] Kim JH, Kim JH, Park JA, Lee SW, Kim WJ, Yu YS, et al. Blood-neural barrier: Intercellular communication at glio-vascular interface. Journal of Biochemistry and Molecular Biology. 2006;**39**(4):339-345

[102] Kumagai AK. Glucose transport in brain and retina: Implications in the management and complications of diabetes. Diabetes/Metabolism Research and Reviews. 1999;**15**(4):261-273

[103] Runkle EA, Antonetti DA. The blood-retinal barrier: Structure and functional significance. Methods in Molecular Biology. 2011;**686**:133-148

[104] Takata K, Hirano H, Kasahara M. Transport of glucose across the blood-tissue barriers. International Review of Cytology. 1997;**172**:1-53

[105] Goncalves A, Ambrosio AF, Fernandes R. Regulation of claudins in blood-tissue barriers under physiological and pathological states. Tissue Barriers. 2013;**1**(3):e24782

[106] Patton N, Aslam TM, MacGillivray T, Deary IJ, Dhillon B, Eikelboom RH, et al. Retinal image analysis: Concepts, applications and potential. Progress in Retinal and Eye Research. 2006;**25**(1):99-127

[107] Diaz-Coranguez M, Ramos C, Antonetti DA. The inner blood-retinal barrier: Cellular basis and development. Vision Research. 2017;**139**:123-137

[108] Lefranc F, Decaestecker C, Brotchi J, Heizmann CW, Dewitte O, Kiss R, et al. Co-expression/co-location of S100 proteins (S100B, S100A1 and S100A2) and protein kinase C (PKC-β, −η and -ζ) in a rat model of cerebral basilar artery vasospasm. Neuropathology and Applied Neurobiology. 2005;**31**(6):649-660

[109] Tiu SC, Chan WY, Heizmann CW, Schäfer BW, Shu SY, Yew DT. Differential expression of S100B1 and S100A61 in the human fetal and aged cerebral cortex. Developmental Brain Research. 2000;**119**(2):159-168

[110] De Bock M, Wang N, Decrock E, Bol M, Gadicherla AK, Culot M, et al. Endothelial calcium dynamics, connexin channels and blood–brain barrier function. Progress in Neurobiology. 2013;**108**:1-20

[111] Kumar P, Shen Q, Pivetti CD, Lee ES, Wu MH, Yuan SY. Molecular mechanisms of endothelial hyperpermeability: Implications in inflammation. Expert Reviews in Molecular Medicine. 2009;**11**:20-e19

[112] Muldoon LL, Alvarez JI, Begley DJ, Boado RJ, Del Zoppo GJ, Doolittle ND, et al. Immunologic privilege in the central nervous system and the blood-brain barrier. Journal of Cerebral Blood Flow and Metabolism. 2013;**33**(1):13-21

[113] Venkataraman V, Nagele RG. Calcium-sensitive ROS-GC1 signaling outside of photoreceptors: A common theme. Molecular and Cellular Biochemistry. 2002;**230**(1-2):117-124

[114] Duda T, Koch KW, Venkataraman V, Lange C, Beyermann M, Sharma RK. Ca^{2+} sensor S100beta-modulated sites of membrane guanylate cyclase in the photoreceptor-bipolar synapse. The EMBO Journal. 2002;**21**(11):2547-2556

[115] Pozdnyakov N, Goraczniak R, Margulis A, Duda T, Sharma RK, Yoshida A, et al. Structural and functional characterization of retinal calcium-dependent guanylate cyclase activator protein (CD-GCAP): Identity with S100beta protein. Biochemistry. 1997;**36**(46):14159-14166

[116] Brozzi F, Arcuri C, Giambanco I, Donato R. S100B protein regulates astrocyte shape and migration via interaction with Src kinase: Implications for astrocyte development, activation, and tumor growth. The Journal of Biological Chemistry. 2009;**284**(13):8797-8811

[117] Zhao Z, Sagare AP, Ma Q, Halliday MR, Kong P, Kisler K, et al. Central role for PICALM in amyloid-beta blood-brain barrier transcytosis and clearance. Nature Neuroscience. 2015;**18**(7):978-987

[118] Aveleira CA, Lin CM, Abcouwer SF, Ambrosio AF, Antonetti DA. TNF-alpha signals through PKCzeta/NF-kappaB to alter the tight junction complex and increase retinal endothelial cell permeability. Diabetes. 2010;**59**(11):2872-2882

[119] Hofmann MA, Drury S, Fu C, Qu W, Taguchi A, Lu Y, et al. RAGE mediates a novel proinflammatory axis: A central cell surface receptor for S100/calgranulin polypeptides. Cell. 1999;**97**(7):889-901

[120] Valencia JV, Mone M, Zhang J, Weetall M, Buxton FP, Hughes TE. Divergent pathways of gene expression are activated by the RAGE ligands S100b and AGE-BSA. Diabetes. 2004;**53**(3):743-751

[121] Goldwaser EL, Acharya NK, Wu H, Godsey GA, Sarkar A, DeMarshall CA, et al. Evidence that brain-reactive autoantibodies contribute to chronic neuronal internalization of exogenous amyloid-beta1-42 and key cell surface proteins during Alzheimer's disease pathogenesis. Journal of Alzheimer's Disease. 2020;**74**:345-361

[122] Nagele EP, Han M, Acharya NK, DeMarshall C, Kosciuk MC, Nagele RG. Natural IgG autoantibodies are abundant and ubiquitous in human sera, and their number is influenced by age, gender, and disease. PLoS One. 2013;**8**(4):e60726

[123] Di Stefano G, Casoli T, Fattoretti P, Gracciotti N, Solazzi M, Bertoni-Freddari C. Distribution of map2 in hippocampus and cerebellum of young and old rats by quantitative immunohistochemistry. The Journal of Histochemistry and Cytochemistry. 2001;**49**(8):1065-1066

[124] Okamoto M, Matsumoto M, Ohtsuki T, Taguchi A, Mikoshiba K, Yanagihara T, et al. Internucleosomal DNA cleavage involved in ischemia-induced neuronal death. Biochemical and Biophysical Research Communications. 1993;**196**(3):1356-1362

[125] Sedeyn JC, Wu H, Hobbs RD, Levin EC, Nagele RG, Venkataraman V. Histamine induces Alzheimer's disease-like blood brain barrier breach and local cellular responses in mouse brain organotypic cultures. BioMed Research International. 2015;**2015**:937148

[126] Wu H, Rodriguez AR, Spur BW, Venkataraman V. An acute retinal model for evaluating blood retinal barrier breach and potential drugs for treatment. Journal of Visualized Experiments. 2016;**115**(e54619):1-8

[127] Levin EC, Acharya NK, Han M, Zavareh SB, Sedeyn JC, Venkataraman V, et al. Brain-reactive autoantibodies are nearly ubiquitous in human sera and may be linked to pathology in the context of blood-brain barrier breakdown. Brain Research. 2010;**1345**:221-232

[128] Arias C, Arrieta I, Massieu L, Tapia R. Neuronal damage and MAP2 changes induced by the glutamate transport inhibitor dihydrokainate and by kainate in rat hippocampus in vivo. Experimental Brain Research. 1997;**116**(3):467-476

[129] D'Andrea MR. Add Alzheimer's disease to the list of autoimmune diseases. Medical Hypotheses. 2005;**64**(3):458-463

[130] DeMarshall C, Sarkar A, Nagele EP, Goldwaser E, Godsey G, Acharya NK, et al. Chapter One—Utility of autoantibodies as biomarkers for diagnosis and staging of neurodegenerative diseases. In: Michael JH, editor. International Review of Neurobiology. Vol. 122. Academic Press; 2015. pp. 1-51

[131] Hughes CG, Morandi A, Girard TD, Riedel B, Thompson JL, Shintani AK, et al. Association between endothelial dysfunction and acute brain dysfunction during critical illness. Anesthesiology. 2013;**118**(3):631-639

[132] Shapiro LA, Whitaker-Azmitia PM. Expression levels of cytoskeletal proteins indicate pathological aging of S100B transgenic mice: An immunohistochemical study of MAP-2, drebrin and GAP-43. Brain Research. 2004;**1019**(1-2):39-46

[133] Acharya NK, Nagele EP, Han M, Nagele RG. Autoantibodies: Double agents in human disease. Science Translational Medicine. 2013;**5**(186):186fs19

[134] D'Andrea MR. Evidence linking neuronal cell death to autoimmunity in Alzheimer's disease. Brain Research. 2003;**982**(1):19-30

[135] Nagele E, Han M, Demarshall C, Belinka B, Nagele R. Diagnosis of Alzheimer's disease based on disease-specific autoantibody profiles in human sera. PLoS One. 2011;**6**(8):e23112

[136] DeMarshall CA, Nagele EP, Sarkar A, Acharya NK, Godsey G, Goldwaser EL, et al. Detection of Alzheimer's disease at mild cognitive impairment and disease progression using autoantibodies as blood-based biomarkers. Alzheimer's &

Dementia (Amsterdam, Netherlands). 2016;**3**:51-62

[137] Bell RD, Zlokovic BV. Neurovascular mechanisms and blood–brain barrier disorder in Alzheimer's disease. Acta Neuropathologica. 2009;**118**(1):103-113

[138] Grammas P. Neurovascular dysfunction, inflammation and endothelial activation: Implications for the pathogenesis of Alzheimer's disease. Journal of Neuroinflammation. 2011;**8**(3):26

[139] D'Andrea MR. Bursting Neurons and Fading Memories: An Alternative Hypothesis of the Pathogenesis of Alzheimer's Disease. Academic Press; 2014

[140] Bouras C, Riederer BM, Kövari E, Hof PR, Giannakopoulos P. Humoral immunity in brain aging and Alzheimer's disease. Brain Research Reviews. 2005;**48**(3):477-487

[141] Stein TD, Fedynyshyn JP, Kalil RE. Circulating autoantibodies recognize and bind dying neurons following injury to the brain. Journal of Neuropathology & Experimental Neurology. 2002;**61**(12):1100-1108

[142] DeMarshall C, Goldwaser EL, Sarkar A, Godsey GA, Acharya NK, Thayasivam U, et al. Autoantibodies as diagnostic biomarkers for the detection and subtyping of multiple sclerosis. Journal of Neuroimmunology. 2017;**309**:51-57

[143] DeMarshall C, Oh E, Kheirkhah R, Sieber F, Zetterberg H, Blennow K, et al. Detection of early-stage Alzheimer's pathology using blood-based autoantibody biomarkers in elderly hip fracture repair patients. PLoS One. 2019;**14**(11):e0225178

[144] Han M, Nagele E, DeMarshall C, Acharya N, Nagele R. Diagnosis of Parkinson's disease based on disease-specific autoantibody profiles in human sera. PLoS One. 2012;**7**(2):e32383

[145] Wen J, Stock AD, Chalmers SA, Putterman C. The role of B cells and autoantibodies in neuropsychiatric lupus. Autoimmunity Reviews. 2016;**15**(9):890-895

[146] Donato R. Intracellular and extracellular roles of S100 proteins. Microscopy Research and Technique. 2003;**60**(6):540-551

[147] Coelho-Aguiar JM, Verissimo CP, Costa DVS, Thomasi BBM, Frauches ACB, Ribiero FP, et al. The enteric glial network acts in the maintenance of intestinal homeostasis and in intestinal disorders. In: Astrocytes. Croatia: IntechOpen; 2019

Chapter 3

The Enteric Glial Network Acts in the Maintenance of Intestinal Homeostasis and in Intestinal Disorders

Juliana de Mattos Coelho-Aguiar, Carla Pires Veríssimo, Deiziane Viana da Silva Costa, Beatriz Bastos de Moraes Thomasi, Ana Carina Bon Frauches, Fabiana Pereira Ribeiro, Ana Lucia Tavares Gomes, Gerly Anne de Castro Brito and Vivaldo Moura-Neto

Abstract

The enteric nervous system (ENS), also known as second brain, innervates our gastrointestinal tract controlling its functions, such as motility, fluid secretion, nutrient absorption, and even involvement in the control of immunity and inflammatory processes. In the gut, the gliocytes are known as enteric glial cells (EGCs). Enteric glial cells form a network that permeates the entire gut. Enteric glia express the cell surface hemichannel of connexin-43 (Cx43) necessary for the propagation of Ca2 + responses, necessary to maintain their functions. In this chapter, besides the development of ENS and its glial cells and the similarities with the astrocytes in the central nervous system, we approached the important role of the glial network in the control of gut homeostasis, in the interaction with the immune system, and its participation in pathological conditions. EGCs are even capable of replacing lost neurons. Thus the enteric glia is a multifunctional cell, which through its multiple interactions maintains the integrity of the ENS allowing it to be resistant to the different and constant aggressions suffered by the digestive system.

Keywords: enteric glial cells, glial network, gut homeostasis, gut inflammation, enteric neurodegeneration

1. Introduction

The enteric nervous system (ENS), also known as second brain, innervates our gastrointestinal tract from the esophagus to the rectum including the pancreas and gallbladder, controlling its functions. The ENS develops from the neural crest cells (NCCs). At the vagal (at the level of somites 1–7) and sacral (posterior to somite 28) regions of the anteroposterior axis, some of NCCs, the enteric neural crest cells (ENCCs), enter the rudimental digestive system, proliferate, and migrate to colonize the primitive gut [1].

The differentiation of enteric neurons starts prior to the enteric glial cells (EGCs). Behind the migratory wavefront, the first neurons arise at E10–E10.5 in the foregut level. Genes of multipotent ENCCs, such as Sox10, FoxD3, and P75, are downregulated, and cells begin to express specific neuronal markers, such as βIIItubulin, RET, HuC/D, and peripherin. Subsequently, at E11.5, the glial differentiation takes place, and the ENCCs downregulate RET expression, while markers such as Sox10, FoxD3, and P75 continue to be expressed. Additionally, other genes that are known to be specifically expressed in EGCs appear, including S100B, glial fibrillary acidic protein (GFAP), and proteolipid protein 1 (PLP1). The development of the mature ENS network is not complete at birth, and the neuronal differentiation is extended to up to 2 weeks after birth (for a detailed description of enteric neurons and glial cells development, [1]).

Initially, ENCCs migrate as intersections of narrow chains of cells. Later on, as development progresses further, these cells aggregate into numerous ganglia that are connected by neuronal projections and EGCs (**Figure 1**). The role of bone morphogenetic proteins (BMPs)-2 and (BMPs)-4 in the neural cell adhesion molecule (NCAM) regulation that are differentially expressed by the cells to form these ganglion-like aggregates is already known [2–6]. The growth factor endothelin3 (EDN3) is important to keep ENS progenitors cells in a proliferative state. It inhibits reversibly the commitment and differentiation of these cells, and in this way it is involved in the correct migration of enteric neural crest cells to colonize the gut [4]. Lack of EDN3 leads to aganglionosis of the distal bowel [5]. It is well known that thyroid hormone 3,5,3′-triiodothyronine (T3) plays an important role in CNS development, and also appears to play a role in the development of the ENS. In vitro, T3 inhibits cell proliferation and stimulated neurite growth of differentiating murine enteric neural crest cells [6]. But, interestingly, this work also showed that spheres of neonate mice ENS progenitor cells increased EDN3 expression by more than 3-fold after T3 treatment, demonstrating a likely crosstalk between these signalling pathways [6]. In the adult mammalian, the ENS is organized into the myenteric and submucosal ganglionated plexuses composed of neurons and EGCs and non-ganglionated plexuses, composed of EGCs that tightly follow neuronal projections that reach all regions of the intestines, including the mucosa. The myenteric plexus (or Auerbach's plexus) is located between the outer longitudinal and circular muscle layers, and the submucosal plexus (or Meissner's plexus) lies in the submucosal region (between the mucosa and the muscular layers) [1].

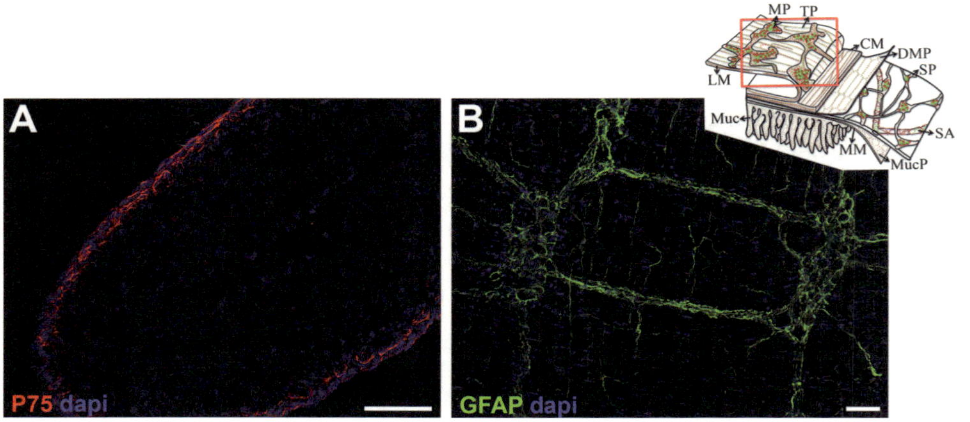

Figure 1.
(A) Transverse section of mouse embryo gut at embryonic day (E)14.5 stained for the glial marker P75. The cells are not yet organized in ganglia. (B) Longitudinal muscle with the adherent myenteric plexus (LMMP) of adult mouse colon. The enteric glial network is evidenced by GFAP staining. Scale bars: 50 μm.

EGCs are distributed across all layers of the intestine and are currently classified into four different subtypes based on their location and morphology. Intraganglionic EGCs (type I) present numerous short and irregular processes and resemble the protoplasmic astrocytes of the central nervous system (CNS); the interganglionic EGCs surround neuronal projections that connect multiple ganglia (type II); the mucosal EGCs (type III) are found around neuronal projections located in the mucosal region and present long and branched processes; and intramuscular EGCs (type IV) are bipolar and elongated and accompany the nerve fibers that cross the muscle layers [7, 8]. In fact, their wide distribution reflects on their performance in different physiological aspects of the gastrointestinal (GI) tract. Indeed, EGCs were shown to participate in the homeostasis of the intestinal epithelial barrier (IEB), to coordinate the GI motility taking part in neurotransmission, and also to modulate inflammation and immune responses.

2. Enteric glia: a unique glial cell type - similarities and differences with astrocytes

In the first studies about enteric glia, the ultrastructure of the glial cell of myenteric plexus was described as a small cell body with many processes. It was suggested that the star-like morphology, as well as the anatomical relation to neurons, resembles astrocytes from the CNS rather than Schwann cells [9]. Jessen et al. [10] showed that intraganglionic EGCs express the characteristic marker of an astrocytic cell, glial fibrillary acidic protein (GFAP), corroborating the assumption that ENS glial cells are analogous to CNS astrocytes [11, 12], although they have different embryological origins.

EGCs and astrocytes exhibit molecular similarities in their electrophysiological properties [13] and express the same group of proteins, including the GFAP [14], and the S100β-linked binding pathway [15]. However, not all properties of the EGCs are similar to astrocytes. They have different embryological origins, for example, astrocytes coming from neuroepithelium and EGCs from neural crest. During the embryonic stages, neuregulin signaling via the ErbB3 receptor is critical for the development of the EGCs, whereas astrocytes do not require such signaling [16]. Unlike astrocytes, EGCs do not express the protein of the aldehyde dehydrogenase 1 L1 (Aldh1L1) [8] but express the transcription factor Sox10 [17] and the protein PLP1, implicated in myelin production and most commonly found in oligodendrocytes and Schwann cells. In fact genic signature of EGCs seems to be more similar to that of oligodendrocytes and Schwann cells than to astrocytes [18].

Similar to astrocytes, EGCs interact with and modulate the performance of different cell types, as we will see throughout this chapter. In addition to interacting with neurons, EGCs establish multidirectional communication with other cell types, such as intestinal mucosal epithelial, muscle, mesenchymal, and immune cells [19].

3. Formation of the gut glial network and the communication through connexin-43 (Cx43) hemichannels

Yet during development, EGCs begin to form a network of interconnected cells that permeate the entire gut (**Figure 1**).

Gabella noted that a striking feature of EGC is the presence of numerous intramembrane particles on its surface [20], and a small part are gap junctions. These intramembrane particles are believed to be hemichannels. It has recently been found that, as astrocytes, enteric glial hemichannels are connexin-43 (Cx43) compounds [21]. Cx43 hemichannels are Ca^{2+}–permeable channels that are also controlled by Ca^{2+} [21].

Like astrocytes, activated EGCs have excitability mediated by transitory intracellular Ca^{2+} elevations, considered central to many functions. Most of the enteric glial receptors for neuroactive compounds are G-protein-coupled receptors, and most of these leads to activation of downstream effectors that elevate intracellular Ca^{2+}. As mentioned by Gulbransen in his book (2014, p. 28) [22], being able to detect the increase in Ca^{2+} levels was essential to establish that neuron-glial communication occurs in ENS and to identify involved mediators. The study realized by McClain et al. [21] also showed the role of Cx43 hemichannels in the propagation of "calcium waves" through the enteric glia network and in the regulation of GI motility [21]. It was shown that glial "calcium waves" activated by extracellular ATP or ADP were disrupted by glial specific loss of Cx43 and result in aberrant ENS network activity and GI dysmotility.

Cx43 expression in EGCs is also related to inflammatory process. Neuronal loss is one of the intestinal inflammation characteristics caused by purinergic receptor activation [23]. Recently, inhibition or genetic ablation of Cx43 in EGCs prevented inflammation-induced neuronal death [24]. This is interesting because it shows that ATP released by EGCs, through Cx43 hemichannels, is involved in both inflammation and motility [21], as mentioned above.

It is possible that Cx43 expression in EGCs is also related to regulation of the intestinal epithelium barrier (IEB). Animals with ablated Cx43 in EGCs also exhibited an increased fluid content in stools [21]; this may imply a role of Cx43 in regulating the IEB, since EGCs have protective effects on enterocytes. A co-culture study showed that EGCs induced in enterocytes an increase in transcription of genes involved in cell-to-cell and cell-to-matrix adhesion and also an increase in cell adhesion [25]. Some of the glia-derived factors, for example, ATP and prostaglandins, could be released through the Cx43 hemichannels [26]. The other effects may come from cell-to-cell contact. EGCs and enterocytes express Cx43 [27], so they may perhaps be joined by Cx43 gap junctions. In fact, the membrane potential of differentiating enterocytes becomes more positive exclusively when they migrate away from the crypt-villous junction [28], possibly due to gap junctions with EGCs (they have higher membrane potential) in this region [29].

In addition to the Cx43 hemichannels, EGCs also have sodium, potassium, and aquaporin-4 channels, whose presence and subtypes vary among their subtypes. Aquaporin channels, for example, are expressed in EGCs within the plexus, but not in extraganglionic EGCs [22].

Even in autism, it has been speculated that inadequate Cx43 expression in EGC could affect GI motility, which is in fact altered in some patients. Some monogenetic autism spectrum disorders are caused by mutations in genes that encode transcriptional or epigenetic factors, for example, methylCpG2-binding protein (MeCP2) in Rett syndrome or TCF4 in Pitt-Hopkins syndrome. These mutations could affect the transcription machinery required for proper expression of Cx43 in EGCs [30].

Thus, EGCs act largely through the release of different molecules, which can happen through the Cx43 hemichannels.

4. Functions of EGCs in gut homeostasis: released factors by EGCs play a role in intestinal epithelial barrier, neurotransmission, and gliotransmission

As already mentioned, EGCs are located throughout intestinal layers and interact with different cell types within the gut. Thus, this cell type is expected to play a number of important roles in the coordination of gut functions. In fact, studies using genetic tools to abrogate GFAP expressing cells resulted in disruption in epithelial integrity, extensive intestinal necrosis, and inflammation, followed by degeneration of enteric neurons [31, 32], evidencing the importance of EGCs for intestinal homeostasis.

EGCs exert their function through the release of important molecules. In the intestine, glial-derived neurotrophic factor (GDNF) is released by EGCs and acts as an anti-apoptotic factor to epithelial cells, neurons, and EGCs [33–36]. GDNF inhibits epithelial cell apoptosis by the activation of GFRα1–GFRα3 receptors and RET co-receptor and the activation of mitogen-activated protein kinase (MAPK) and phosphatidylinositol-3-kinase/serine–threonine kinase (PI3K/AKT) signaling pathways [37, 38]. GDNF also inhibits apoptosis of EGCs in an autocrine manner [34]. Mu opioid receptor activation by morphine in EGCs decreases their GDNF synthesis, with consequent IEB disruption [39]. Moreover, GDNF has been shown to increase the integrity of the IEB via ZO-1 upregulation [38].

Among neuroactive molecules, ATP is the most well-characterized molecule released from EGCs. ATP is released through the opening of Cx 43 hemichannels [21, 40]. More specifically, the released ATP modifies adjacent glia, triggering intercellular Ca^{2+} waves and influencing adjacent neurons. The ATP released by EGCs may induce neuronal cell death, as discussed above (topic 2). Nitric oxide (NO), a key inhibitory neurotransmitter in the ENS and a factor that drives oxidative stress in disease, is also produced by EGCs [41, 42]. This molecule is produced by the enzyme inducible nitric oxide synthase (iNOS). Under pathological stimuli, EGCs express iNOS and produce large amounts of NO, which may be protective or deleterious depending on the circumstances. Some evidence, however, suggest that EGCs may constitutively express iNOS and that NO plays a physiological role by modulating epithelial ion transport [41].

The data above have suggested that EGCs play an essential role in neuronal support and neurotransmission. Indeed, EGCs actively participate in neurotransmission. Intraganglionic EGCs provide enteric neurons with essential precursors for the synthesis of neurotransmitters such as NO [43, 44], glutamate, and γ-aminobutyric acid (GABA) [45]. In health, EGCs provide antioxidants, like reduced glutathione [46, 47], and growth factors (e.g., GDNF) [48] to neurons. In addition, EGCs support neurotransmission by regulating the bioavailability of neuroactive substances in the extracellular environment. EGC enzymes are essential for the removal of neuroactive compounds surrounding enteric neurons. Moreover, glial potassium channels maintain neurotransmission and prevent the death of excitotoxic neurons by regulating and buffering potassium [13, 49].

A poorly understood question is how EGCs interpret and process the signals they receive from enteric neurons to eventually play their presumptive roles in neurons and GI function.

Little is known about EGCs activating ligands, but the Ca^{2+} transients probably trigger different modes of gliotransmission, such as Ca^{2+}–dependent exocytosis or factor release through the Cx43 hemichannels [50].

Boesmans et al. [51] demonstrated that enteric neurons can communicate with adjacent EGCs, releasing purines through their panexin channels. In fact, ATP and purines are the most ubiquitous signaling molecules involved in the enteric transmission of neurons to glia in vitro [52, 53] and in situ [21, 23, 54–56], but EGCs also have other receptors that allow glia to initiate responses to neurotransmitters released by neurons and other neuroactive substances, including receptors to norepinephrine, glutamate, thrombin, lipid signaling molecules, serotonin, bradykinin, histamine, and endothelin [22].

As mentioned before, it is already known that in vitro propagation of Ca^{2+} responses between EGCs depends on ATP release through hemichannels [40]. And later it was seen that substances released by Cx43 hemichannels mediate intercellular communication between EGCs [21].

How other populations of EGCs outside the enteric ganglia interact with enteric neurons is currently unknown.

5. EGC plasticity

The ENS has a significant ability to adapt to microenvironmental influences throughout life, either by inflammatory bowel diseases or by changes in eating habits [57]. The mechanisms of cellular communication involved in the plasticity of EGCs are not yet fully understood. Understanding how EGCs act and especially how they perform the role of progenitor cell and differentiate into neuron is of paramount importance for a better understanding of how the ENS performs its complex functions.

It has already been shown that numerous neural crest-derived stem cells are found in different locations in the adult organism, including the intestine [58]. When isolated by flow cytometry for p75 markers [59] or integrin-α2 [60], or also through dissociation for in vitro cultivation of neurospheres [61, 62], EGCs can give rise to a large number of other cells including glial cells, neurons, and even myofibroblasts. Following transplantation of these cells into intestinal explants, EGCs differentiate into glial and neuronal cells [61, 63]. These data underscore the plastic potential of EGCs that, when transplanted into the CNS, are able to function as oligodendrocytes and astrocytes [26]. It is noteworthy that under different physiological conditions and after injury [60], EGCs proliferate and differentiate into neuron only upon specific injury situations [64, 65]. Liu and colleagues have shown that it is possible to induce neurogenesis in the myenteric plexus in vivo by activating the serotonin receptor upon administration of the 5-hydroxytryptamine 4 (5-HT 4) agonist [66]. Indeed, other studies in postnatal bowel suggest that serotonin also promotes ENS repair and neurogenesis via 5-HT4 receptor [67, 68]. Moreover, studies have shown that mouse and human EGCs undergo neurogenesis after colitis [65, 69].

Under chemical injury with benzalkonium chloride detergent (BAC), it was possible to observe neurogenesis in vivo. About 3 months after injury, EGCs adjacent to the aganglionic area give rise to sox10-positive glial cells expressing the neuronal marker HuC/D [64]. It has been proposed that interruption of contact between cells (ganglion structure dissociation) may initiate neurogenesis from precursor cells expressing EGCs markers (sox10, p75, S100β, GFAP). These studies suggest that tissue dissociation to establish cell culture, as well as that observed in chemical injury, could activate the neurogenic potential of EGCs.

A recent work, however, suggested that constitutive neurogenesis occurs in the gut [70], contrasting with data obtained by other groups that suggest that intestinal neurons are not easily replaced under healthy conditions [71–73]. Moreover, this study highlighted a population of nestin-positive adult progenitor cells that give rise to new neurons, different from that of GFAP-positive EGCs. These data contrast with previous works that had shown nestin and GFAP co-expression by EGCs [60]. Furthermore, nestin-expressing intestinal NSCs cells give rise to neurosphere-derived neurons and glia in vitro. Besides these cells can differentiate into glial, neuronal, and mesenchymal lineages in vitro and also generate neurons in vivo [74].

EGCs do not produce extracellular matrix (ECM). However, their processes contact basal lamina proteins including heparan sulfate proteoglycan, type IV collagen, and laminin [11, 75, 76]. This suggests that the microenvironment is also a factor of great relevance for the function of EGCs and neurons. Recent data demonstrate that EGCs in vitro, in absence of appropriate substrates were stimulated to initiate neuronal differentiation. Therefore, it seems that the contact of adult EGCs with laminin plays a crucial role in inhibiting their potential for neuronal differentiation (Veríssimo et al., 2009).

6. Implications of EGCs in pathological conditions

The importance of the correct neuron-glial communication is evidenced in a situation of intestinal inflammation and neurodegeneration, when EGCs act as a direct mediator of neuronal cell death.

6.1 EGCs in inflammatory bowel diseases (IBD)

Chronic inflammation in the GI tract can cause important changes in the ENS, as demonstrated by several studies in patients with IBD, such as ulcerative colitis (UC) and Crohn's disease (CD) [77, 78]. Both UC and CD are characterized by inflammation, which is accompanied by the release of a range of pro-inflammatory cytokines, following intestinal dysmotility [79, 80].

An increase in GDNF and GFAP immunolabeling was observed in EGCs in inflamed colonic mucosa of patients with UC, CD, and *Clostridioides difficile* (*C. difficile*) infectious colitis [78]. In addition, S100B upregulation has also been identified in a variety of diseases, such as UC [42, 81, 82], celiac disease [83], and intestinal mucositis induced by antineoplastic drugs [84, 85]. Increased expression of S100B was also found in the intestine of humans with *C. difficile* infection (CDI), in animal model of CDI, and in mouse ileal loop injected with *C. difficile* toxin A (TcdA) (unpublished data).

In intestinal injury, reactive gliosis is a response of EGCs to protect the neuronal network during intestinal inflammation [86]. However, depending on the degree of inflammation, this event may cause damage to neurons and to EGCs themselves due to a deregulated response of these cells to the virulence factors of pathogenic bacteria or pro-inflammatory mediators released by immune cells, neurons, or EGCs. This dual effects may have an important effect in the instability of the release of protective and dangerous factors, such as GDNF, an anti-apoptotic factor, and S100B, a pro-inflammatory cytokine, by EGCs [34, 35].

6.1.1 Ulcerative colitis and Crohn's disease

A strong upregulation in levels of GDNF was reported in the intestinal crypts and in the myenteric and submucosal plexuses in patients with CD. In UC, GDNF immunoreactivity was reported to be less pronounced than CD. However, no alteration in GFR-1 was evidenced in patients with CD and UC. GFR-1 is a receptor for GDNF binding that is predominantly found at the basolateral parts of the human colonic epithelium [37].

The EGCs regulate the epithelial barrier function and inflammation through the release of S-nitrosoglutathione (GSNO), a potent nitric oxide donor. Interestingly, EGCs are the main source of GSNO within the intestine. It has been shown that the levels of GSNO are reduced in CD and UC [35]. GSNO regulates the intestinal permeability by stimulating in enterocytes the upregulation of proteins of tight junctions, such as occludins and ZO-1, and inhibiting the increase of phosphorylated myosin light chain (PMLC), as well as improving the location of these proteins [87].

EGCs from human colonic tissues with Crohn's disease have reduced 15-hydroxyeicosatetraenoic (15-HETE) synthesis. As GSNO, 15-HETE controls the paracellular permeability of the IEB by inhibiting adenosine monophosphate-activated protein kinase (AMPK) and regulating ZO-1 expression [88].

NO, such as tumor necrosis factor-α (*TNF-α) and PGE2, plays an important role in the pathogenesis of ulcerative colitis and is secreted by EGCs.* S100B can induce increased NO release, as well as TNF-α and PGE2, by murine and human EGCs via S100B/TLR4 [42, 82].

Losses in 61% of the enteric neurons and 38% of the EGCs have been reported during ulcerative colitis in human [77]. In fact, activation of EGCs during colitis induced by dinitrobenzene sulfonic acid (DNBS) in mice had been shown to be the central mechanism in the development of enteric neuropathy, since the gating of glial Cx43 hemichannels by nitric oxide and subsequent ATP release are required for enteric neuron death [24].

6.1.2 Colitis by Clostridioides difficile

C. difficile is an obligate anaerobic, spore-forming Gram-positive bacillus that can colonize, germinate, and proliferate in the human gut after antibiotic use [89, 90]. The incidence of *C. difficile* infection (CDI) across the world has increased with 107,760 admissions per year [91, 92]. The clinical disease ranges from mild diarrhea to toxic megacolon, colonic perforation, and death [93].

The major virulence factors of *C. difficile* are toxins A and B (TcdA and TcdB). TcdA and TcdB stimulate the release of a variety of mediators such as interleukin (IL)-1β, IL-17, IL-23, TNF-α, CXC motif chemokine ligand 4 (CXCL4), CXCL2, and inhibitory macrophage migration factor (MIF) in several cells, such as epithelial cells, immune cells, and enteric neurons [94–97]. In contrast to those cells, EGCs challenged with TcdA and TcdB do not release detectable levels of IL-1β, interferon-gamma (INF-γ), and TNF-α [98].

The first studies on changes in the ENS evoked by *C. difficile* toxins showed that TcdA and TcdB excite enteric neurons stimulating the release of substance P and vasoactive intestinal peptide (VIP) via noradrenergic transmission inhibition and IL-1β pathway, respectively, resulting in neutrophil recruitment and secretory diarrhea [99–101].

A recent study demonstrated increased cell population expressing both HuC/D and SOX2 in inflamed colonic tissues in patients with CDI [65]. So, these EGCs are important for generating new neurons after intestinal injury.

A study of 447 and 444 patients with *C. difficile*-associated diarrhea acquired in the community and hospital, respectively, showed GI dysmotility in these patients after 1 year of the last diarrhea episode. Among the dysmotilities are IBD, gastro-esophageal reflux disease, constipation, and dyspepsia [102]. These dysmotilities have been shown to be related to ENS changes. Deregulated activation of EGCs during inflammation may alter their regulatory role in motility (discussed in topic 3), causing intestinal dysmotility.

It was demonstrated that TcdB stimulates morphological alteration and apoptosis in EGCs *in vitro* [98, 103]. TcdB-induced apoptosis of EGCs involves the NADPH oxidase/ROS/JNK/caspase-3 signaling pathway independently of the mitochondrial pathway [103]. In addition, TcdB induces senescence in EGCs. Cell senescence is characterized by alterations in the cell cycle, changes in metabolism, morphology, and gene expression that together may contribute to persistent inflammation [104].

6.1.3 Inflammation by other causes

It has been demonstrated that EGCs are involved in decreased infection foci and IL-8 secretion and in the inhibition of alterations in IEB resistance in infection by

Shigella flexneri in rabbit ileal loop model. Similar findings were found in human colonic mucosa explants infected with *S. flexneri*. GSNO released by EGCs showed to be a mediator responsible for protecting epithelial cells from *S. flexneri*-induced effects [105].

Giardia duodenalis (also known as *G. lamblia* or *G. intestinalis*) is a protozoan parasite capable of causing sporadic or epidemic diarrheal illness. Giardia duodenalis-induced infection is one of the most common human parasitic diseases worldwide [106]. Studies have shown a reduction in EGCs from the submucosal plexus of the mouse duodenum and jejunum during infection induced by assemblages A and B of *G. duodenalis*. However, only assemblages B of *G. duodenalis were observed to* induce a reduction in those cells from the duodenal myenteric plexus. Surprisingly, mice infected with *G. duodenalis* did not exhibit diarrhea and any alterations in GI transit time [107].

Antineoplastic drugs, such as 5-FU, irinotecan, and oxaliplatin, have been currently used to treat several types of cancer, including breast and colorectal cancer. Mucositis and diarrhea are common side effects of these antineoplastic drugs [108]. Many cells are stimulated to release inflammatory mediators during intestinal mucositis, and persistent GI over-contractility has also been demonstrated, even after inflammation has resolved, suggesting that chemotherapy might affect gut neuronal and EGC function [109].

During intestinal mucositis induced by oxaliplatin, reduced GFAP and increased S100B protein expression were evidenced, as well as reduced co-localization of GFAP and S100B in ileal myenteric plexus of mice [110].

In fact, increased S100B release by EGCs has been shown to be a mediator in charge of causing neuronal death, as well as reactive gliosis, epithelial damage, and inflammatory response (release of IL-6, TNFα, and NO) during 5-FU-induced intestinal mucositis via S100B/RAGE/NFκB [84].

As we will deepen later, EGCs can be stimulated by immune cells during intestinal inflammation. A recent study showed that mediators released by mast cells cause reactive gliosis and neuronal death together with the intestinal mucositis induced by irinotecan [85].

Figure 2 shows a schematic highlighting how EGCs are affected by and participates on intestinal inflammation.

6.2 EGCs in neurodegenerative diseases

Due to its great interaction with neurons and modulation of neuronal responses, it is possible to imagine that EGCs play a central role in neurodegenerative diseases. Indeed, the role of EGCs seems to be compromised in many neurodegenerative diseases, and this is true for both CNS and ENS.

Enteric neurodegeneration is a common marker for a group of diseases classically known as enteric neuropathies. The changes found present as alterations in enteric smooth cells and/or compromised functioning of the ENS—often impacting in GI motility [111]. The neuropathies chronic intestinal pseudo-obstruction (CIPO) and slow transit constipation (STC) are characterized by neurodegeneration affecting the lower GI tract. Moreover, it has already been shown that enteric glia is implicated in Parkinson's disease (PD), and participation in Alzheimer's disease (AD) is speculated [111].

CIPO is a condition characterized by failure of GI motility without apparent mechanical lesion [112]. Histological patterns show different classes of the disease depending on the cell type involvement (enteric neurons, smooth muscle cells, and interstitial cells of Cajal). Enteric neuron degeneration promotes intestinal neuromuscular disorders [111, 113]. In chronic idiopathic intestinal pseudo-obstruction

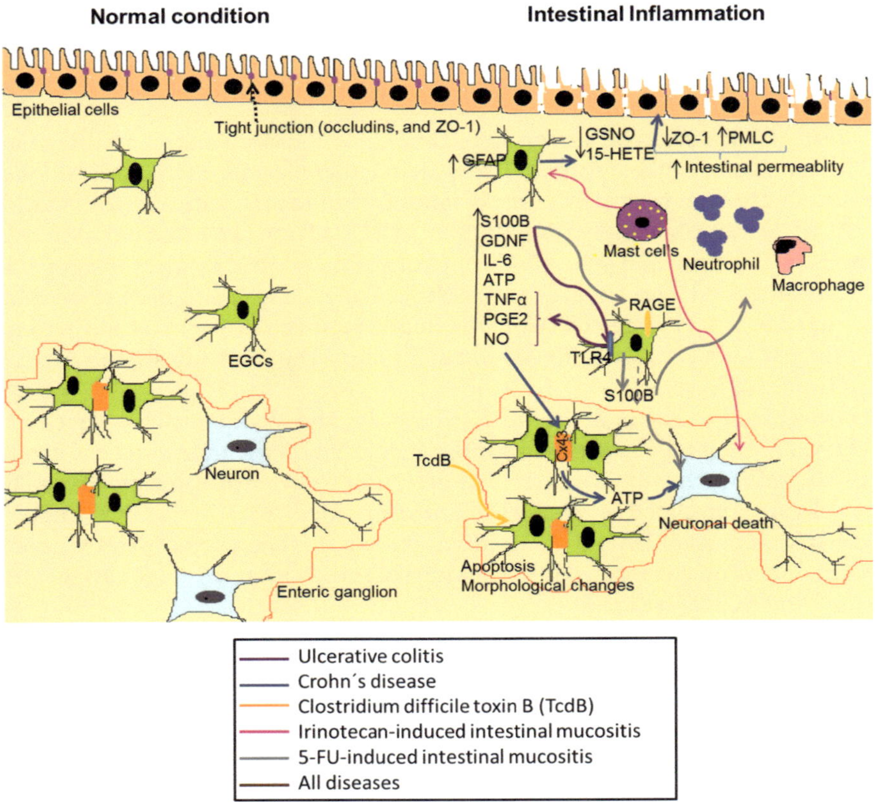

Figure 2.
Mediator release by EGCs during intestinal inflammation and their role in the pathogenesis of intestinal inflammatory diseases. During intestinal inflammation promoted by ulcerative colitis, Crohn's disease, colitis induced by C. difficile, *irinotecan- and 5-FU-induced intestinal mucositis, and enteric glial factors (S100B, GDNF and GFAP) are upregulated. EGCs are stimulated to secrete S100B, GDNF, IL-6, ATP, TNF-α, PGE2, and NO. In addition, EGCs produce reduced levels of GSNO and 15-HETE during Crohn's disease, resulting in increase of intestinal permeability by ZO-1 downregulation and PMLC upregulation. NO release by EGCs promotes Cx43 opening with consequent ATP release by EGCs, resulting in neuronal death in Crohn's disease. In 5-FU-induced intestinal mucositis, S100B released by EGCs via RAGE receptor activation drives reactive gliosis, neuronal death, and immune cell activation, whereas mediators released by mast cells induce reactive gliosis and neuronal death during irinotecan-induced intestinal mucositis . In ulcerative colitis, S100B activates TLR4, resulting in TNFα, PGE2, and NO by EGCs.* C. difficile *Toxin B (TcdB) induces EGCs apoptosis and morphological changes.*

(CIIP), EGC infection by JC virus (polyomavirus) has been described suggesting a role of enteric glia in this enteric neuropathy [114].

Constipation is a common functional GI disorder characterized by infrequent bowel motions and/or incomplete defecation [115]. Studies on the neuronal subtypes involved in the STC pathogenesis are still very uncertain. It was pointed that excessive production of NO in the colonic myenteric plexus of STC patients would inhibit propulsive contraction. Results about other neurotransmitters as VIP, substance P, and serotonin were contradictory [113]. Besides the decrease of enteric neurons and interstitial cells of Cajal, STC also presents a significant decrease of EGCs [116], and some discussion has emerged about constipation being a neuro-gliopathy [79]. Several reports showed that different conditions presenting constipation have a feature: loss of EGCs, and it points to a pathophysiological meaning since the EGC directly regulates enteric neurons and interstitial cells of Cajal through neurotrophic factors [116, 117] and ATP signaling [79, 118].

6.2.1 Parkinson's disease

In the last years, the literature has shown that some pathological conditions, such as PD, classically described to compromise the CNS are now recognized as multicentric neurodegenerative processes since they affect different systems such as the ENS [119–122]. A number of non-motor symptoms in PD have been identified, and many of them manifest early, even before the clinical stage of the disease (characterized by emergence of the classic motor features) when the diagnosis can be made [123]. They found lesions in autopsies of patients by identifying the presence of intraneuronal inclusions called Lewy bodies/neuritis, which are described as protein agglomerates where α-synuclein is the main constituent. The areas primarily affected were olfactory structures, the dorsal motor nucleus of the vagus nerve and the ENS [124, 125]. According to Braak's hypothesis, there could occur a migration of the ENS lesion via the vagus nerve to the CNS [124]. Indeed, Lewy neurites are detectable in the presymptomatic stage of PD along the autonomic pathways and in the GI tract [126]. Besides, analysis of human colon biopsies obtained 2 to 5 years before PD onset showed the presence of pathologic α-synuclein in neurodegeneration sites, suggesting that colonic α-synuclein staining can be considered a biomarker of premotor PD symptoms [127].

GI symptoms are the most debilitating PD non-motor features and are present in almost every patient at some stage of the disorder [124, 128, 129]. The symptoms commonly reported by patients are weight loss, dysphagia, decreased frequency of intestinal peristalsis, and difficulty in defecation [130]. Recent evidences indicate that PD pathological alterations in the gut involve EGCs and probably impairment of their critical role in GI physiology. In fact, colonic biopsies of PD patients showed an increased expression of GFAP both at the transcript and protein levels [131, 132] as well as a reduction in GFAP phosphorylation. These features strongly suggest that reactive gliosis may be associated with degenerative diseases [131].

In PD colon biopsies the upregulation of GFAP was accompanied by an increase in the expression of pro-inflammatory cytokines, mainly TNF-α, IFN-γ, IL-1β, and IL-6 [132]. These data suggest a link between glial dysfunction and enteric inflammation in the colon of PD patients.

Alterations in IEB have been observed in patients and animal models of PD [133, 134]. Modifications in protein levels and protein distribution that compose the barrier (e.g., occludin) were documented [128]. In agreement with this, PD patients show fecal biomarkers of inflammation as calprotectin and also increased intestinal permeability as alpha-1-antitrypsin [134]. It is known that the IEB is strongly regulated by EGCs [37–39]. Since EGCs is sensitized in PD patients and modulates all these processes, it is speculated that IEB could be impaired by altered glial signaling which could contribute to the inflammatory process.

Intestinal dysmotility is the symptom that affects directly patient's quality of life and is shared among PD patients. Constipation is the most common non-motor symptom manifested in both prodromal and clinic phases of PD [135–137]. Recently, constipation was included as a criterion for prodromal PD diagnostic, and discussion about the validation of constipation as a risk factor for the development of PD has been recurrent [138]. As already discussed, impairments in EGCs activity produce constipation due to a loss in the neural control of gut motility [21].

However, despite the evidence, there is still no direct demonstration of how enteric glia is involved in PD, either in the cause of the disease or its consequences.

In this way, the suggestion that PD could onset in the gut emerges from the identification of activated EGCs, local inflammation, impaired IEB, aggregation of α-synuclein in neurons, and GI disorders in a window prior to the appearance of classic motor deficits. Recently, Seguella et al. suggest that EGCs could be the "missing link" that connects the ENS to the CNS [139]. The authors called attention to enteric glial cell-mediated inflammatory response, which could reach the CNS by the gut-brain axis and lead to neuronal cell death and disruption of synaptic interactions [139, 140]. Thus, EGCs would function as an "entrance door" to noxious stimuli from the intestinal lumen that could damage the CNS. However, the mechanisms by which the pro-inflammatory glial mediators rise to the CNS still remains to be clarified [139].

6.2.2 Alzheimer's disease

Alzheimer's disease (AD) is the most common neurodegenerative disorder affecting people in the world. The neurodegeneration causes a progressive cognitive decline and loss of working memory [141]. Among the non-cognitive symptoms of AD are the GI symptoms which point to a role of ENS in AD [142]. In fact, the brain biomarker of AD, the extracellular plaques containing β-amyloid, has already been described in the intestinal submucosa of patients [143] which is in agreement with the expression of amyloid precursor protein in enteric neurons and also EGCs [144]. The discussion of the peripheral immune response has been widely debated as the pathogenic pattern of AD that contributes to central neurodegeneration [145, 146]. In this context, some discussion has been raised about EGC possibly acting as a peripheric coordinator of immune differentiation of T cells [139] since EGCs express the major histocompatibility complex II and T-cell costimulatory molecules [147–149]. As mentioned above, EGCs are able to respond to an inflammatory environment contributing to the process, activate enteric neurodegenerative mechanisms, and immunomodulate the IEB. All these features could contribute to an inflammatory peripheral state and sensibilization of CNS through the blood–brain barrier [139]. It is still speculative to relate these glial interactions to AD, but there are indications of an immunomodulatory relevance of this cell in the GIT, as will be discussed again below.

6.3 Interactions between EGCs and the immune system

Recently, insightful and essential findings have shed light in the field of neuroimmunology, especially with the development of high solution technological approaches to underlie neuroimmune communications. It has been proposed that the immune and the nervous systems interact in health and disease and are expected to function alongside to promote tissue homeostasis [150]. More specifically, neuroimmune interactions have been suggested by understanding the relative anatomical positioning of cell types and their dynamics within the tissue in homeostasis and response to insults. Moreover, the expression of corresponding ligands and receptors by immune and nervous cells, for instance, may determine physiological interactions between the two systems. However, efforts to identify mechanisms to decipher how immune and neural cells interact in a steady-state environment and respond to genetic and epigenetic cues are still a challenge to be addressed.

Because the GI tract is the connection between the external with the internal environments of the body, the ENS is continuously exposed and expected to interact with the extrinsic (dietary and microbiota-derived metabolites) and intrinsic (immune system and stromal cells) environments of the gut. The strategical anatomical positioning of ENS and immune cells throughout the GI wall and their physiological features are crucial to defeat pathogens and maintain the intestinal homeostasis. Emerging studies have identified two distinct types of tissue-resident macrophages within the intestinal wall that are closely associated with ENS cells [150]. Lamina propria macrophages (LpMs) preferentially display pro-inflammatory phenotype and are the most abundant cell group located just beneath the intestinal epithelium. These cells, together with neuronal processes and mucosal EGCs, form tight physical and functional barriers that protect the intestines against pathogens, although the mechanisms that underlie those interactions are still to be further explored [150].

At the level of the myenteric plexus, muscularis macrophages (MMs) are closely associated with neuron cell bodies and fibers and EGCs and present a tissue-protective phenotype. Similar to microglia in the CNS, MMs can phagocyte neuronal debris during homeostasis [70]. Another population of gut self-maintaining macrophages (gMacs) was described to be fundamental for ENS homeostasis since the genetic depletion of those macrophages led to a loss of enteric neurons resulting in reduced intestinal function [151]. Moreover, enteric neurons and innate lymphoid cells type 2 (ILC2) functionally integrate to initiate type 2 immune responses. The integration between neuron-ILC2 units is necessary for cytokine production and inflammation repair upon worm infection [152, 153].

EGCs also appear to participate in immune responses, but so far, its impact on immune cells is still relatively unexplored under homeostasis. However, an exciting study has recently discussed that GDNF secreted by EGCs activates IL-22-producing ILC3 via Ret signaling [154]. Interestingly, Ret signaling regulates Peyer's patches organogenesis, underlining the prospective role of EGCs in orchestrating innate immune functions in the gut [155]. Furthermore, experiments performed in the submucosal plexus (SMP) from patients with functional dyspepsia (FD) showed that morphological alterations both in EGCs and neurons are due to increased numbers of eosinophil and mast cell within ganglionic structures [156]. This also suggests that EGCs and the immune system work together to maintain the intestinal homeostasis.

It is known that EGCs protect T lymphocytes from cell death by upregulating the expression of IL-7 after exposure to pro-inflammatory cytokines such as IL-1β and TNFα [157]. Moreover, EGCs were suggested to have immunosuppressive characteristics in CD by inhibiting T-cell proliferation [158]). Nonetheless, the cellular and molecular mechanisms that govern the role of EGCs in intestinal pathologies remain unclear. EGCs express MHCII [148] that is upregulated under inflammatory conditions [149, 159], conferring an immunological feature to these cells in a pathological environment. Moreover, EGCs can secrete and respond to IL-1β, IL-6, and IL-10 and nitric oxide in vitro, as already mentioned, suggesting another property of these cells in the mediation of inflammatory responses [24, 160, 161]. Although, it is plausible that EGCs have an important function in modulating neuroimmune interactions, understanding their specific contributions to the maintenance of the gut homeostasis would be useful to decipher their roles in inflammatory disorders.

This would possibly suggest an immune protective role of EGCs to maintain the mucosal barrier. However, those studies failed in showing direct evidence that EGCs are necessary for intestinal barrier function. On the other hand, studies in which EGCs were disrupted but not entirely ablated did not show any noticeable signs of inflammation. In contrast, disruption in EGC homeostasis culminated in changes in mucosal function as well as in neurochemical coding, leading to alterations in enteric neurons and consequently in motor activity [162–164]. Thus, taken together, the immunological roles of EGCs protecting the intestinal environment from damage remain contentious by using genetic tools to ablate/disturb these cells.

7. Conclusion

As we could notice in this chapter, there are still many unexplained aspects of the EGCs physiology. Although we have already found interesting studies that show their relation with neurons or alterations in cases of inflammation, the exact mechanisms by which EGCs activates neurons to control GI motility are still unknown. Little is known about their interaction with the immune system, for example, or their participation in neurodegenerative diseases that affect both ENS and CNS. Recently, Seguella et al. suggest that EGCs could be the "missing link" that connects the ENS to the CNS [139]. EGCs in the context of disease could be an important target for diagnosis and therapy of many intestinal and neurological disorders.

Taken together, these evidence show the importance of EGCs for the maintenance of intestinal homeostasis and that disturbance of glial functions could alter GI physiology through the modulation of neurotransmission and of the responses of the different cellular types or even activation of cellular signals to enter the neuronal differentiation processes in specific situations.

Author details

Juliana de Mattos Coelho-Aguiar[1,2*], Carla Pires Veríssimo[2], Deiziane Viana da Silva Costa[3], Beatriz Bastos de Moraes Thomasi[4], Ana Carina Bon Frauches[5], Fabiana Pereira Ribeiro[2], Ana Lucia Tavares Gomes[4], Gerly Anne de Castro Brito[3] and Vivaldo Moura-Neto[1,2]

1 Instituto Estadual do Cérebro Paulo Niemeyer, Secretaria de Estado de Saúde do Rio de Janeiro, Rio de Janeiro, Brazil

2 Instituto de Ciências Biomédicas, Universidade Federal do Rio de Janeiro, Rio de Janeiro, Brazil

3 Departamento de Morfologia, Faculdade de Medicina, Universidade Federal do Ceará, Fortaleza, Ceará, Brazil

4 Departamento de Neurobiologia, Universidade Federal Fluminense, Niterói, Rio de Janeiro, Brazil

5 The Francis Crick Institute, London, United Kingdom

*Address all correspondence to: jumcoelho@gmail.com

References

[1] Coelho-Aguiar J, Bon-Frauches AC, Gomes ALT, Verissimo CP, Aguiar DP, Matias D, et al. The enteric glia: Identity and functions. Glia. 2015;**63**(6):921-935

[2] Faure C, Chalazonitis A, Rheaume C, Bouchard G, Sampathkumar S-G, Yarema KJ, et al. Gangliogenesis in the enteric nervous system: Roles of the polysialylation of the neural cell adhesion molecule and its regulation by bone morphogenetic protein-4. Developmental Dynamics: An Official Publication of the American Association of the Anatomists. 2007 Jan;**236**(1):44-59

[3] Rollo BN, Zhang D, Simkin JE, Menheniott TR, Newgreen DF. Why are enteric ganglia so small? Role of differential adhesion of enteric neurons and enteric neural crest cells. F1000Research. 2015;**4**:113

[4] Bondurand N, Natarajan D, Barlow A, Thapar N, Pachnis V. Maintenance of mammalian enteric nervous system progenitors by SOX10 and endothelin 3 signalling. Development. 2006;**133**:2075-2086

[5] Woodward MN, Sidebotham EL, Connell MG, Kenny SE, Vaillant CR, Lloyd DA, et al. Analysis of the effects of endothelin-3 on the development of neural crest cells in the embryonic mouse gut. Journal of Pediatric Surgery. 2003;**38**:1322-1328

[6] Mohr R, Neckel P, Zhang Y, Stachon S, Nothelfer K, Schaeferhoff K, et al. Molecular and cell biological effects of 3,5,3'-triiodothyronine on progenitor cells of the enteric nervous system in vitro. Stem Cell Research. 2013;**11**:1191-1205

[7] Gulbransen BD, Sharkey KA. Novel functional roles for enteric glia in the gastrointestinal tract. Nature Reviews. Gastroenterology & Hepatology [Internet]. 2012;**9**(11):625-632. Available from: http://www.ncbi.nlm.nih.gov/pubmed/22890111

[8] Boesmans W, Lasrado R, Vanden Berghe P, Pachnis V. Heterogeneity and phenotypic plasticity of glial cells in the mammalian enteric nervous system. Glia [Internet]. 2015;**63**(2):229-241. Available from: http://www.ncbi.nlm.nih.gov/pubmed/25161129

[9] Gabella G. Glial cells in the myenteric plexus. Zeitschrift fur Naturforschung Teil B, Chemie, Biochemie, Biophysik, Biologie und verwandte Gebiete. 1971;**26**(3):244-245

[10] Jessen KR, Duance VC, Mirsky R, Timpl R, Bannerman PGC. Light microscopic immunolocalization of laminin, type IV collagen, nidogen, heparan sulphate proteoglycan and fibronectin in the enteric nervous system of rat and Guinea pig. Journal of Neurocytology. 2005;**15**(6):733-743

[11] Gershon MD, Rothman TP. Enteric glia. Glia. 1991;**4**(2):195-204

[12] Jessen KR, Mirsky R. The origin and development of glial cells in peripheral nerves. Nature Reviews. Neuroscience. 2005;**6**(9):671-682

[13] Hanani M, Francke M, Hartig W, Grosche J, Reichenbach A, Pannicke T. Patch-clamp study of neurons and glial cells in isolated myenteric ganglia. American Journal of Physiology. Gastrointestinal and Liver Physiology. 2000;**278**(4):G644-G651

[14] Jessen KR, Mirsky R. Glial cells in the enteric nervous system contain glial fibrillary acidic protein. Nature [Internet]. 1980;**286**(5774):736-737. Available from: http://www.ncbi.nlm.nih.gov/pubmed/6997753

[15] Ferri GL, Probert L, Cocchia D, Michetti F, Marangos PJ, Polak JM. Evidence for the presence of S-100 protein in the glial component of the human enteric nervous system. Nature [Internet]. 1982;**297**(5865):409-410. Available from: http://www.ncbi.nlm.nih.gov/pubmed/7043279

[16] Riethmacher D, Sonnenberg-Riethmacher E, Brinkmann V, Yamaai T, Lewin GR, Birchmeier C. Severe neuropathies in mice with targeted mutations in the ErbB3 receptor. Nature. 1997;**389**(6652):725-730

[17] Young HM, Jones BR, McKeown SJ. The projections of early enteric neurons are influenced by the direction of neural crest cell migration. The Journal of Neuroscience: The Official Journal of the Society for Neuroscience. 2002;**22**(14):6005-6018

[18] Rao M, Nelms BD, Dong L, Salinas-Rios V, Rutlin M, Gershon MD, et al. Enteric glia express proteolipid protein 1 and are a transcriptionally unique population of glia in the mammalian nervous system. Glia [Internet]. 2015;**63**(11):2040-2057. Available from: http://www.ncbi.nlm.nih.gov/pubmed/26119414

[19] Ruhl A. Glial cells in the gut. Neurogastroenterology and Motility [Internet]. 2005;**17**(6):777-790. Available from: http://www.ncbi.nlm.nih.gov/pubmed/16336493

[20] Gabella G. Ultrastructure of the nerve plexuses of the mammalian intestine: The enteric glial cells. Neuroscience [Internet]. 1981;**6**(3):425-436. Available from: http://www.ncbi.nlm.nih.gov/pubmed/7219723

[21] McClain JL, Grubisic V, Fried D, Gomez-Suarez RA, Leinninger GM, Sevigny J, et al. Ca^{2+} responses in enteric glia are mediated by connexin-43 hemichannels and modulate colonic transit in mice. Gastroenterology [Internet]. 2014;**146**(2):497-507. Available from: http://www.ncbi.nlm.nih.gov/pubmed/24211490

[22] Gulbransen BD. Enteric glia. Colloquium Series on Neuroglia in Biology and Medicine: From Physiology to Disease [Internet]. 2014;**1**(2):1-70. Available from: http://www.morganclaypool.com/doi/10.4199/C00113ED1V01Y201407NGL002

[23] Gulbransen BD, Bashashati M, Hirota SA, Gui X, Roberts JA, MacDonald JA, et al. Activation of neuronal P2X7 receptor-pannexin-1 mediates death of enteric neurons during colitis. Nature Medicine. 2012;**18**(4):600-604

[24] Brown IA, McClain JL, Watson RE, Patel BA, Gulbransen BD. Enteric glia mediate neuron death in colitis through purinergic pathways that require connexin-43 and nitric oxide. Cellular and Molecular Gastroenterology and Hepatology [Internet]. 2016;**2**(1):77-91. Available from: http://www.ncbi.nlm.nih.gov/pubmed/26771001

[25] Van Landeghem L, Mahe MM, Teusan R, Leger J, Guisle I, Houlgatte R, et al. Regulation of intestinal epithelial cells transcriptome by enteric glial cells: Impact on intestinal epithelial barrier functions. BMC Genomics [Internet]. 2009;**10**:507. Available from: http://www.ncbi.nlm.nih.gov/pubmed/19883504

[26] Cherian PP, Siller-Jackson AJ, Gu S, Wang X, Bonewald LF, Sprague E, et al. Mechanical strain opens connexin 43 hemichannels in osteocytes: A novel mechanism for the release of prostaglandin. Molecular Biology of the Cell. 2005;**16**(7):3100-3106

[27] Leaphart CL, Qureshi F, Cetin S, Li J, Dubowski T, Baty C, et al. Interferon-gamma inhibits intestinal

restitution by preventing gap junction communication between enterocytes. Gastroenterology. 2007;**132**(7):2395-2411

[28] Cremaschi D, James PS, Meyer G, Smith MW. Positional dependence of enterocyte membrane potential in hamster and rabbit enterocytes. Comparative Biochemistry and Physiology. A, Comparative Physiology. 1984;**78**(4):661-666

[29] Liu YA, Chung YC, Pan ST, Shen MY, Hou YC, Peng SJ, et al. 3-D imaging, illustration, and quantitation of enteric glial network in transparent human colon mucosa. Neurogastroenterology and Motility: The Official Journal of the European Gastrointestinal Motility Society. 2013;**25**(5):e324-e338

[30] Grubisic V, Parpura V. The second brain in autism spectrum disorder: Could connexin 43 expressed in enteric glial cells play a role? Frontiers in Cellular Neuroscience. 2015;**9**:242

[31] Bush TG, Savidge TC, Freeman TC, Cox HJ, Campbell EA, Mucke L, et al. Fulminant jejuno-ileitis following ablation of enteric glia in adult transgenic mice. Cell [Internet]. 1998;**93**(2):189-201. Available from: http://www.ncbi.nlm.nih.gov/pubmed/9568712

[32] Cornet A, Savidge TC, Cabarrocas J, Deng WL, Colombel JF, Lassmann H, et al. Enterocolitis induced by autoimmune targeting of enteric glial cells: A possible mechanism in Crohn's disease? Proceedings of the National Academy of Sciences of the United States of America [Internet]. 2001;**98**(23):13306-13311. Available from: http://www.ncbi.nlm.nih.gov/pubmed/11687633

[33] Wang N, Li K, Song S, Chen J. Gastric electrical stimulation improves enteric neuronal survival. International Journal of Molecular Medicine [Internet]. 2017;**40**(2):438-446. Available from: https://www.ncbi.nlm.nih.gov/pubmed/28627584

[34] Steinkamp M, Gundel H, Schulte N, Spaniol U, Pflueger C, Zizer E, et al. GDNF protects enteric glia from apoptosis: Evidence for an autocrine loop. BMC Gastroenterology [Internet]. 2012;**12**:6. Available from: https://www.ncbi.nlm.nih.gov/pubmed/22251670

[35] Savidge TC, Newman P, Pothoulakis C, Ruhl A, Neunlist M, Bourreille A, et al. Enteric glia regulate intestinal barrier function and inflammation via release of S-nitrosoglutathione. Gastroenterology [Internet]. 2007;**132**(4):1344-1358. Available from: http://www.ncbi.nlm.nih.gov/pubmed/17408650

[36] Schäfer KH, Mestres P. The GDNF-induced neurite outgrowth and neuronal survival in dissociated myenteric plexus cultures of the rat small intestine decreases postnatally. Experimental Brain Research [Internet]. 1999;**125**(4):447-452. Available from: https://www.ncbi.nlm.nih.gov/pubmed/10323291

[37] Steinkamp M, Geerling I, Seufferlein T, von Boyen G, Egger B, Grossmann J, et al. Glial-derived neurotrophic factor regulates apoptosis in colonic epithelial cells. Gastroenterology [Internet]. 2003;**124**(7):1748-1757. Available from: http://www.ncbi.nlm.nih.gov/pubmed/12806607

[38] Zhang DK, He FQ, Li TK, Pang XH, Cui DJ, Xie Q, et al. Glial-derived neurotrophic factor regulates intestinal epithelial barrier function and inflammation and is therapeutic for murine colitis. The Journal of Pathology [Internet]. 2010;**222**(2):213-222. Available from: http://www.ncbi.nlm.nih.gov/pubmed/20632386

[39] Bauman BD, Meng J, Zhang L, Louiselle A, Zheng E, Banerjee S, et al. Enteric glial-mediated enhancement of intestinal barrier integrity is compromised by morphine. The Journal of Surgical Research [Internet]. 2017;**219**:214-221. Available from: https://www.ncbi.nlm.nih.gov/pubmed/29078884

[40] Zhang W, Segura BJ, Lin TR, Hu Y, Mulholland MW. Intercellular calcium waves in cultured enteric glia from neonatal Guinea pig. Glia. 2003;**42**(3):252-262

[41] MacEachern SJ, Patel BA, McKay DM, Sharkey KA. Nitric oxide regulation of colonic epithelial ion transport: A novel role for enteric glia in the myenteric plexus. The Journal of Physiology. 2011;**589**(13):3333-3348

[42] Esposito G, Capoccia E, Turco F, Palumbo I, Lu J, Steardo A, et al. Palmitoylethanolamide improves colon inflammation through an enteric glia/toll like receptor 4-dependent PPAR-alpha activation. Gut [Internet]. 2014;**63**(8):1300-1312. Available from: http://www.ncbi.nlm.nih.gov/pubmed/24082036

[43] Aoki E, Semba R, Kashiwamata S. Evidence for the presence of L-arginine in the glial components of the peripheral nervous system. Brain Research. 1991;**559**(1):159-162

[44] Nagahama M, Semba R, Tsuzuki M, Aoki E. L-arginine immunoreactive enteric glial cells in the enteric nervous system of rat ileum. Biological Signals and Receptors [Internet]. 2001;**10**(5):336-340. Available from: http://www.ncbi.nlm.nih.gov/pubmed/11490099

[45] Jessen KR, Mirsky R. Astrocyte-like glia in the peripheral nervous system: An immunohistochemical study of enteric glia. The Journal of Neuroscience [Internet]. 1983;**3**(11):2206-2218. Available from: http://www.ncbi.nlm.nih.gov/pubmed/6138397

[46] Abdo H, Mahe MM, Derkinderen P, Bach-Ngohou K, Neunlist M, Lardeux B. The omega-6 fatty acid derivative 15-deoxy-delta(1)(2),(1)(4)-prostaglandin J2 is involved in neuroprotection by enteric glial cells against oxidative stress. The Journal of Physiology [Internet]. 2012;**590**(11):2739-2750. Available from: http://www.ncbi.nlm.nih.gov/pubmed/22473776

[47] Abdo H, Derkinderen P, Gomes P, Chevalier J, Aubert P, Masson D, et al. Enteric glial cells protect neurons from oxidative stress in part via reduced glutathione. The FASEB Journal [Internet]. 2010;**24**(4):1082-1094. Available from: http://www.ncbi.nlm.nih.gov/pubmed/19906678

[48] von Boyen GB, Steinkamp M, Reinshagen M, Schafer KH, Adler G, Kirsch J. Nerve growth factor secretion in cultured enteric glia cells is modulated by proinflammatory cytokines. Journal of Neuroendocrinology [Internet]. 2006;**18**(11):820-825. Available from: http://www.ncbi.nlm.nih.gov/pubmed/17026531

[49] Costagliola A, Van Nassauw L, Snyders D, Adriaensen D, Timmermans J-P. Voltage-gated delayed rectifier K v 1-subunits may serve as distinctive markers for enteroglial cells with different phenotypes in the murine ileum. Neuroscience Letters. 2009;**461**(2):80-84

[50] Grubisic V, Parpura V. Two modes of enteric gliotransmission differentially affect gut physiology. Glia. 2017;**65**(5):699-711

[51] Boesmans W, Ameloot K, van den Abbeel V, Tack J, Vanden Berghe P. Cannabinoid receptor 1 signalling dampens activity

and mitochondrial transport in networks of enteric neurones. Neurogastroenterology and Motility: The Official Journal of the European Gastrointestinal Motility Society. 2009;**21**(9):958-e77

[52] Kimball BC, Mulholland MW. Enteric glia exhibit P2U receptors that increase cytosolic calcium by a phospholipase C-dependent mechanism. Journal of Neurochemistry [Internet]. 1996;**66**(2):604-612. Available from: http://www.ncbi.nlm.nih.gov/pubmed/8592130

[53] Gomes P, Chevalier J, Boesmans W, Roosen L, van den Abbeel V, Neunlist M, et al. ATP-dependent paracrine communication between enteric neurons and glia in a primary cell culture derived from embryonic mice. Neurogastroenterology and Motility: The Official Journal of the European Gastrointestinal Motility Society. 2009;**21**(8):870-e62

[54] Gulbransen BD, Sharkey KA. Purinergic neuron-to-glia signaling in the enteric nervous system. Gastroenterology [Internet]. 2009;**136**(4):1349-1358. Available from: http://www.ncbi.nlm.nih.gov/pubmed/19250649

[55] Gulbransen BD, Bains JS, Sharkey KA. Enteric glia are targets of the sympathetic innervation of the myenteric plexus in the Guinea pig distal colon. The Journal of Neuroscience: The Official Journal of the Society for Neuroscience. 2010;**30**(19):6801-6809

[56] Broadhead MJ, Bayguinov PO, Okamoto T, Heredia DJ, Smith TK. Ca^{2+} transients in myenteric glial cells during the colonic migrating motor complex in the isolated murine large intestine. The Journal of Physiology [Internet]. 2012;**590**(2):335-350. Available from: http://www.ncbi.nlm.nih.gov/pubmed/22063626

[57] Schafer KH, Ginneken C, Copray S. Plasticity and neural stem cells in the enteric nervous system. Anatomical Record [Internet]. 2009;**292**(12):1940-1952. Available from: http://www.ncbi.nlm.nih.gov/entrez/query.fcgi?cmd=Retrieve&db=PubMed&dopt=Citation&list_uids=19943347

[58] Dupin E, Coelho-Aguiar JM. Isolation and differentiation properties of neural crest stem cells. Cytometry Part A: The Journal of the International Society for Analytical Cytology. 2013;**83**(1):38-47

[59] Kruger GM, Mosher JT, Bixby S, Joseph N, Iwashita T, Morrison SJ. Neural crest stem cells persist in the adult gut but undergo changes in self-renewal, neuronal subtype potential, and factor responsiveness. Neuron. 2002;**35**(4):657-669

[60] Joseph NM, He S, Quintana E, Kim Y-G, Nunez G, Morrison SJ. Enteric glia are multipotent in culture but primarily form glia in the adult rodent gut. The Journal of Clinical Investigation. 2011 Sep;**121**(9):3398-3411

[61] Almond S, Lindley RM, Kenny SE, Connell MG, Edgar DH. Characterisation and transplantation of enteric nervous system progenitor cells. Gut. 2007;**56**(4):489-496

[62] Bondurand N, Natarajan D, Thapar N, Atkins C, Pachnis V. Neuron and glia generating progenitors of the mammalian enteric nervous system isolated from foetal and postnatal gut cultures. Development (Cambridge, England). 2003 Dec;**130**(25):6387-6400

[63] Rauch U, Hansgen A, Hagl C, Holland-Cunz S, Schafer K-H. Isolation and cultivation of neuronal precursor cells from the developing human enteric nervous system as a tool for cell therapy in dysganglionosis. International Journal of Colorectal Disease. 2006;**21**(6):554-559

[64] Laranjeira C, Sandgren K, Kessaris N, Richardson W, Potocnik A, Vanden Berghe P, et al. Glial cells in the mouse enteric nervous system can undergo neurogenesis in response to injury. The Journal of Clinical Investigation. 2011 Sep;**121**(9):3412-3424

[65] Belkind-Gerson J, Graham HK, Reynolds J, Hotta R, Nagy N, Cheng L, et al. Colitis promotes neuronal differentiation of Sox2+ and PLP1+ enteric cells. Scientific Reports [Internet]. 2017;**7**(1):2525. Available from: https://www.ncbi.nlm.nih.gov/pubmed/28566702

[66] Liu M-T, Kuan Y-H, Wang J, Hen R, Gershon MD. 5-HT4 receptor-mediated neuroprotection and neurogenesis in the enteric nervous system of adult mice. The Journal of Neuroscience: The Official Journal of the Society for Neuroscience. 2009 Aug;**29**(31):9683-9699

[67] Gershon MD. Serotonin is a sword and a shield of the bowel: Serotonin plays offense and defense. Transactions of the American Clinical and Climatological Association. 2012;**123**:268-280

[68] Matsuyoshi H, Kuniyasu H, Okumura M, Misawa H, Katsui R, Zhang G-X, et al. A 5-HT(4)-receptor activation-induced neural plasticity enhances in vivo reconstructs of enteric nerve circuit insult. Neurogastroenterology and Motility: The Official Journal of the European Gastrointestinal Motility Society. 2010;**22**(7):806-813

[69] Belkind-Gerson J, Hotta R, Nagy N, Thomas AR, Graham H, Cheng L, et al. Colitis induces enteric neurogenesis through a 5-HT4-dependent mechanism. Inflammatory Bowel Diseases. 2015;**21**(4):870-878

[70] Kulkarni S, Micci M-A, Leser J, Shin C, Tang S-C, Fu Y-Y, et al. Adult enteric nervous system in health is maintained by a dynamic balance between neuronal apoptosis and neurogenesis. Proceedings of the National Academy of Sciences of the United States of America. 2017;**114**(18):E3709-E3718

[71] Pham TD, Gershon MD, Rothman TP. Time of origin of neurons in the murine enteric nervous system: Sequence in relation to phenotype. The Journal of Comparative Neurology. 1991;**314**(4):789-798

[72] Sasselli V, Pachnis V, Burns AJ. The enteric nervous system. Developmental Biology. 2012 Jun;**366**(1):64-73

[73] Young HM, Bergner AJ, Muller T. Acquisition of neuronal and glial markers by neural crest-derived cells in the mouse intestine. The Journal of Comparative Neurology. 2003 Jan;**456**(1):1-11

[74] Belkind-Gerson J, Carreon-Rodriguez A, Benedict LA, Steiger C, Pieretti A, Nagy N, et al. Nestin-expressing cells in the gut give rise to enteric neurons and glial cells. Neurogastroenterology and Motility: The Official Journal of the European Gastrointestinal Motility Society. 2013;**25**(1):61-69

[75] Bannerman PG, Mirsky R, Jessen KR, Timpl R, Duance VC. Light microscopic immunolocalization of laminin, type IV collagen, nidogen, heparan sulphate proteoglycan and fibronectin in the enteric nervous system of rat and Guinea pig. Journal of Neurocytology. 1986;**15**(6):733-743

[76] Neunlist M, Aubert P, Bonnaud S, Van Landeghem L, Coron E, Wedel T, et al. Enteric glia inhibit intestinal epithelial cell proliferation partly through a TGF-beta1-dependent pathway. American Journal of Physiology. Gastrointestinal and Liver Physiology. 2007;**292**(1):G231-G241

[77] Bernardini N, Segnani C, Ippolito C, De Giorgio R, Colucci R, Faussone-Pellegrini MS, et al. Immunohistochemical analysis of myenteric ganglia and interstitial cells of Cajal in ulcerative colitis. Journal of Cellular and Molecular Medicine [Internet]. 2012;**16**(2):318-327. Available from: http://www.ncbi.nlm.nih.gov/pubmed/21426484

[78] von Boyen GB, Schulte N, Pfluger C, Spaniol U, Hartmann C, Steinkamp M. Distribution of enteric glia and GDNF during gut inflammation. BMC Gastroenterology [Internet]. 2011;**11**:3. Available from: http://www.ncbi.nlm.nih.gov/pubmed/21235736

[79] Bassotti G, Villanacci V, Rostami Nejad M. Chronic constipation: no more idiopathic, but a true neuropathological entity. Gastroenterology and Hepatology from Bed to Bench [Internet]. 2011;**4**(3):109-115. Available from: http://www.ncbi.nlm.nih.gov/pubmed/24834167

[80] Schreiber S, Nikolaus S, Hampe J, Hämling J, Koop I, Groessner B, et al. Tumour necrosis factor alpha and interleukin 1beta in relapse of Crohn's disease. Lancet [Internet]. 1999;**353**(9151):459-461. Available from: https://www.ncbi.nlm.nih.gov/pubmed/9989717

[81] Celikbilek A, Celikbilek M, Sabah S, Tanık N, Borekci E, Dogan S, et al. The serum S100B level as a biomarker of enteroglial activation in patients with ulcerative colitis. International Journal of Inflammation [Internet]. 2014;**2014**:986525. Available from: https://www.ncbi.nlm.nih.gov/pubmed/24790767

[82] Cirillo C, Sarnelli G, Esposito G, Grosso M, Petruzzelli R, Izzo P, et al. Increased mucosal nitric oxide production in ulcerative colitis is mediated in part by the enteroglial-derived S100B protein. Neurogastroenterology and Motility [Internet]. 2009;**21**(11):1209-e112. Available from: https://www.ncbi.nlm.nih.gov/pubmed/19558426

[83] Esposito G, Cirillo C, Sarnelli G, De Filippis D, D'Armiento FP, Rocco A, et al. Enteric glial-derived S100B protein stimulates nitric oxide production in celiac disease. Gastroenterology [Internet]. 2007;**133**(3):918-925. Available from: http://www.ncbi.nlm.nih.gov/pubmed/17678654

[84] Costa DVS, Bon-Frauches AC, Silva AMHP, Lima-Júnior RCP, Martins CS, Leitão RFC, et al. 5-fluorouracil induces enteric neuron death and glial activation during intestinal mucositis via a S100B-RAGE-NFκB-dependent pathway. Scientific Reports [Internet]. 2019;**9**(1):665. Available from: https://www.ncbi.nlm.nih.gov/pubmed/30679569

[85] Nogueira LT, Costa DV, Gomes AS, Martins CS, Silva AM, Coelho-Aguiar JM, et al. The involvement of mast cells in the irinotecan-induced enteric neurons loss and reactive gliosis. Journal of Neuroinflammation [Internet]. 2017;**14**(1):79. Available from: https://www.ncbi.nlm.nih.gov/pubmed/28388962

[86] Burda JE, Sofroniew MV. Reactive gliosis and the multicellular response to CNS damage and disease. Neuron [Internet]. 2014;**81**(2):229-248. Available from: http://www.ncbi.nlm.nih.gov/pubmed/24462092

[87] Cheadle GA, Costantini TW, Bansal V, Eliceiri BP, Coimbra R. Cholinergic signaling in the gut: A novel mechanism of barrier protection through activation of enteric glia cells. Surgical Infections [Internet]. 2014;**15**(4):387-393. Available from: http://www.ncbi.nlm.nih.gov/pubmed/24828283

[88] Pochard C, Coquenlorge S, Jaulin J, Cenac N, Vergnolle N, Meurette G, et al.

Defects in 15-HETE production and control of epithelial permeability by human enteric glial cells from patients with Crohn's disease. Gastroenterology [Internet]. 2016;**150**(1):168-180. Available from: https://www.ncbi.nlm.nih.gov/pubmed/26433161

[89] Hopkins RJ, Wilson RB. Treatment of recurrent. Gastroenterology Report [Internet]. 2018;**6**(1):21-28. Available from: https://www.ncbi.nlm.nih.gov/pubmed/29479439

[90] Martin JS, Monaghan TM, Wilcox MH. Clostridium difficile infection: Epidemiology, diagnosis and understanding transmission. Nature Reviews. Gastroenterology & Hepatology [Internet]. 2016;**13**(4):206-216. Available from: https://www.ncbi.nlm.nih.gov/pubmed/26956066

[91] Peery AF, Crockett SD, Murphy CC, Lund JL, Dellon ES, Williams JL, et al. Burden and cost of gastrointestinal, liver, and pancreatic diseases in the United States: Update 2018. Gastroenterology [Internet]. 2019;**156**(1):254-272. Available from: https://www.ncbi.nlm.nih.gov/pubmed/30315778

[92] Davies KA, Longshaw CM, Davis GL, Bouza E, Barbut F, Barna Z, et al. Underdiagnosis of *Clostridium difficile* across Europe: The European, multicentre, prospective, biannual, point-prevalence study of *Clostridium difficile* infection in hospitalised patients with diarrhoea (EUCLID). The Lancet Infectious Diseases [Internet]. 2014;**14**(12):1208-1219. Available from: https://www.ncbi.nlm.nih.gov/pubmed/25455988

[93] Walker AS, Eyre DW, Wyllie DH, Dingle KE, Griffiths D, Shine B, et al. Relationship between bacterial strain type, host biomarkers, and mortality in *Clostridium difficile* infection. Clinical Infectious Diseases [Internet]. 2013;**56**(11):1589-1600. Available from: https://www.ncbi.nlm.nih.gov/pubmed/23463640

[94] Jose S, Mukherjee A, Abhyankar MM, Leng L, Bucala R, Sharma D, et al. Neutralization of macrophage migration inhibitory factor improves host survival after *Clostridium difficile* infection. Anaerobe [Internet]. 2018;**53**:56-63. Available from: https://www.ncbi.nlm.nih.gov/pubmed/29944928

[95] Yu H, Chen K, Sun Y, Carter M, Garey KW, Savidge TC, et al. Cytokines are markers of the *Clostridium difficile*-induced inflammatory response and predict disease severity. Clinical and Vaccine Immunology [Internet]. 2017;**24**(8):e00037-17. Available from: https://www.ncbi.nlm.nih.gov/pubmed/28592627

[96] Nakagawa T, Mori N, Kajiwara C, Kimura S, Akasaka Y, Ishii Y, et al. Endogenous IL-17 as a factor determining the severity of *Clostridium difficile* infection in mice. Journal of Medical Microbiology [Internet]. 2016;**65**(8):821-827. Available from: https://www.ncbi.nlm.nih.gov/pubmed/27166143

[97] Buonomo EL, Madan R, Pramoonjago P, Li L, Okusa MD, Petri WA. Role of interleukin 23 signaling in *Clostridium difficile* colitis. The Journal of Infectious Diseases [Internet]. 2013;**208**(6):917-920. Available from: https://www.ncbi.nlm.nih.gov/pubmed/23776194

[98] Fettucciari K, Ponsini P, Gioè D, Macchioni L, Palumbo C, Antonelli E, et al. Enteric glial cells are susceptible to *Clostridium difficile* toxin B. Cellular and Molecular Life Sciences [Internet]. 2017;**74**(8):1527-1551. Available from: https://www.ncbi.nlm.nih.gov/pubmed/27891552

[99] Neunlist M, Barouk J, Michel K, Just I, Oreshkova T, Schemann M, et al. Toxin B of *Clostridium difficile* activates human VIP submucosal neurons, in part via an IL-1beta-dependent pathway. American Journal of Physiology.

Gastrointestinal and Liver Physiology [Internet]. 2003;**285**(5):G1049-G1055. Available from: https://www.ncbi.nlm.nih.gov/pubmed/12801886

[100] Xia Y, Hu HZ, Liu S, Pothoulakis C, Wood JD. Clostridium difficile toxin A excites enteric neurones and suppresses sympathetic neurotransmission in the Guinea pig. Gut [Internet]. 2000;**46**(4):481-486. Available from: https://www.ncbi.nlm.nih.gov/pubmed/10716676

[101] Pothoulakis C, Castagliuolo I, LaMont JT, Jaffer A, O'Keane JC, Snider RM, et al. CP-96,345, a substance P antagonist, inhibits rat intestinal responses to *Clostridium difficile* toxin A but not cholera toxin. Proceedings of the National Academy of Sciences of the United States of America [Internet]. 1994;**91**(3):947-951. Available from: https://www.ncbi.nlm.nih.gov/pubmed/7508124

[102] Gutiérrez RL, Riddle MS, Porter CK. Increased risk of functional gastrointestinal sequelae after *Clostridium difficile* infection among active duty United States military personnel (1998-2010). Gastroenterology [Internet]. 2015;**149**(6):1408-1414. Available from: https://www.ncbi.nlm.nih.gov/pubmed/26255560

[103] Macchioni L, Davidescu M, Fettucciari K, Petricciuolo M, Gatticchi L, Gioè D, et al. Enteric glial cells counteract *Clostridium difficile* toxin B through a NADPH oxidase/ROS/JNK/caspase-3 axis, without involving mitochondrial pathways. Scientific Reports [Internet]. 2017;7:45569. Available from: https://www.ncbi.nlm.nih.gov/pubmed/28349972

[104] Fettucciari K, Macchioni L, Davidescu M, Scarpelli P, Palumbo C, Corazzi L, et al. *Clostridium difficile* toxin B induces senescence in enteric glial cells: A potential new mechanism of *Clostridium difficile* pathogenesis. Biochimica et Biophysica Acta, Molecular Cell Research [Internet]. 2018;**1865**(12):1945-1958. Available from: https://www.ncbi.nlm.nih.gov/pubmed/30296499

[105] Flamant M, Aubert P, Rolli-Derkinderen M, Bourreille A, Neunlist MR, Mahé MM, et al. Enteric glia protect against *Shigella flexneri* invasion in intestinal epithelial cells: A role for S-nitrosoglutathione. Gut [Internet]. 2011;**60**(4):473-484. Available from: https://www.ncbi.nlm.nih.gov/pubmed/21139062

[106] El Basha NR, Zaki MM, Hassanin OM, Rehan MK, Omran D. Giardia assemblages a and B in diarrheic patients: A comparative study in Egyptian children and adults. The Journal of Parasitology [Internet]. 2016;**102**(1):69-74. Available from: https://www.ncbi.nlm.nih.gov/pubmed/26509291

[107] Pavanelli MF, Colli CM, Bezagio RC, Góis MB, de Melo G, de Almeida Araújo EJ, et al. Assemblages A and B of *Giardia duodenalis* reduce enteric glial cells in the small intestine in mice. Parasitology Research [Internet]. 2018;**117**(7):2025-2033. Available from: https://www.ncbi.nlm.nih.gov/pubmed/29728828

[108] Soveri LM, Hermunen K, de Gramont A, Poussa T, Quinaux E, Bono P, et al. Association of adverse events and survival in colorectal cancer patients treated with adjuvant 5-fluorouracil and leucovorin: Is efficacy an impact of toxicity? European Journal of Cancer [Internet]. 2014;**50**(17):2966-2974. Available from: https://www.ncbi.nlm.nih.gov/pubmed/25266443

[109] Soares PM, Mota JM, Gomes AS, Oliveira RB, Assreuy AM, Brito GA, et al. Gastrointestinal dysmotility in 5-fluorouracil-induced intestinal mucositis outlasts inflammatory process

resolution. Cancer Chemotherapy and Pharmacology [Internet]. 2008;**63**(1):91-98. Available from: https://www.ncbi.nlm.nih.gov/pubmed/18324404

[110] Robinson AM, Stojanovska V, Rahman AA, McQuade RM, Senior PV, Nurgali K. Effects of oxaliplatin treatment on the enteric glial cells and neurons in the mouse ileum. The Journal of Histochemistry and Cytochemistry [Internet]. 2016;**64**(9):530-545. Available from: https://www.ncbi.nlm.nih.gov/pubmed/27389702

[111] Knowles CH, Lindberg G, Panza E, De Giorgio R. New perspectives in the diagnosis and management of enteric neuropathies. Nature Reviews. Gastroenterology & Hepatology. 2013;**10**(4):206-218

[112] Cogliandro RF, De Giorgio R, Barbara G, Cogliandro L, Concordia A, Corinaldesi R, et al. Chronic intestinal pseudo-obstruction. Best Practice & Research. Clinical Gastroenterology. 2007;**21**(4):657-669

[113] Goldstein AM, Thapar N, Karunaratne TB, De Giorgio R. Clinical aspects of neurointestinal disease: Pathophysiology, diagnosis, and treatment. Developmental Biology. 2016 Sep;**417**(2):217-228

[114] Selgrad M, De Giorgio R, Fini L, Cogliandro RF, Williams S, Stanghellini V, et al. JC virus infects the enteric glia of patients with chronic idiopathic intestinal pseudo-obstruction. Gut. 2009 Jan;**58**(1):25-32

[115] Tabbers MM, DiLorenzo C, Berger MY, Faure C, Langendam MW, Nurko S, et al. Evaluation and treatment of functional constipation in infants and children: Evidence-based recommendations from ESPGHAN and NASPGHAN. Journal of Pediatric Gastroenterology and Nutrition. 2014 Feb;**58**(2):258-274

[116] Bassotti G, Villanacci V, Maurer CA, Fisogni S, Di Fabio F, Cadei M, et al. The role of glial cells and apoptosis of enteric neurones in the neuropathology of intractable slow transit constipation. Gut [Internet]. 2006;**55**(1):41-46. Available from: http://www.ncbi.nlm.nih.gov/pubmed/16041063

[117] Bassotti G, Villanacci V, Antonelli E, Morelli A, Salerni B. Enteric glial cells: New players in gastrointestinal motility? Laboratory Investigation [Internet]. 2007;**87**(7):628-632. Available from: http://www.ncbi.nlm.nih.gov/pubmed/17483847

[118] Burnstock G, Lavin S. Interstitial cells of Cajal and purinergic signalling. Autonomic Neuroscience : Basic & Clinical. 2002 Apr;**97**(1):68-72

[119] Braak H, Del Tredici K. Invited article: Nervous system pathology in sporadic Parkinson disease. Neurology [Internet]. 2008;**70**(20):1916-1925. Available from: http://www.ncbi.nlm.nih.gov/pubmed/18474848

[120] Chalazonitis A, Rao M. Enteric nervous system manifestations of neurodegenerative disease. Brain Research [Internet]. 2018;**1693**(Pt B):207-213. Available from: http://www.ncbi.nlm.nih.gov/entrez/query.fcgi?cmd=Retrieve&db=PubMed&dopt=Citation&list_uids=29360466

[121] Lebouvier T, Chaumette T, Paillusson S, Duyckaerts C, Bruley d, Varannes S, et al. The second brain and Parkinson's disease. The European Journal of Neuroscience [Internet]. 2009;**30**(5):735-741. Available from: http://www.ncbi.nlm.nih.gov/pubmed/19712093

[122] Natale G, Pasquali L, Paparelli A, Fornai F. Parallel manifestations of neuropathologies in the enteric and central nervous systems.

Neurogastroenterology and Motility [Internet]. 2011;**23**(12):1056-1065. Available from: http://www.ncbi.nlm.nih.gov/entrez/query.fcgi?cmd=Retrieve&db=PubMed&dopt=Citation&list_uids=21951862

[123] Winkler J, Ehret R, Buttner T, Dillmann U, Fogel W, Sabolek M, et al. Parkinson's disease risk score: Moving to a premotor diagnosis. Journal of Neurology [Internet]. 2011;**258**(Suppl 2):S311-S315. Available from: http://www.ncbi.nlm.nih.gov/pubmed/21560061

[124] Braak H, Del Tredici K, Rub U, de Vos RA, Jansen Steur EN, Braak E. Staging of brain pathology related to sporadic Parkinson's disease. Neurobiology of Aging [Internet]. 2003;**24**(2):197-211. Available from: http://www.ncbi.nlm.nih.gov/pubmed/12498954

[125] Braak H, de Vos RA, Bohl J, Del Tredici K. Gastric alpha-synuclein immunoreactive inclusions in Meissner's and Auerbach's plexuses in cases staged for Parkinson's disease-related brain pathology. Neuroscience Letters [Internet]. 2006;**396**(1):67-72. Available from: http://www.ncbi.nlm.nih.gov/pubmed/16330147

[126] Lebouvier T, Neunlist M, Bruley des, Varannes S, Coron E, Drouard A, N'Guyen JM, et al. Colonic biopsies to assess the neuropathology of Parkinson's disease and its relationship with symptoms. PLoS One [Internet]. 2010;**5**(9):e12728. Available from: http://www.ncbi.nlm.nih.gov/pubmed/20856865

[127] Shannon KM, Keshavarzian A, Mutlu E, Dodiya HB, Daian D, Jaglin JA, et al. Alpha-synuclein in colonic submucosa in early untreated Parkinson's disease. Movement Disorders [Internet]. 2012;**27**(6):709-715. Available from: http://www.ncbi.nlm.nih.gov/pubmed/21766334

[128] Clairembault T, Leclair-Visonneau L, Coron E, Bourreille A, Le Dily S, Vavasseur F, et al. Structural alterations of the intestinal epithelial barrier in Parkinson's disease. Acta Neuropathologica Communications [Internet]. 2015;**3**:12. Available from: http://www.ncbi.nlm.nih.gov/pubmed/25775153

[129] Dickson DW, Fujishiro H, Orr C, DelleDonne A, Josephs KA, Frigerio R, et al. Neuropathology of non-motor features of Parkinson disease. Parkinsonism & Related Disorders [Internet]. 2009;**15**(Suppl 3):S1-S5. Available from: http://www.ncbi.nlm.nih.gov/entrez/query.fcgi?cmd=Retrieve&db=PubMed&dopt=Citation&list_uids=20082965

[130] Pfeiffer RF. Gastrointestinal dysfunction in Parkinson's disease. Parkinsonism & Related Disorders [Internet]. 2011;**17**(1):10-15. Available from: http://www.ncbi.nlm.nih.gov/pubmed/20829091

[131] Clairembault T, Kamphuis W, Leclair-Visonneau L, Rolli-Derkinderen M, Coron E, Neunlist M, et al. Enteric GFAP expression and phosphorylation in Parkinson's disease. Journal of Neurochemistry [Internet]. 2014;**130**(6):805-815. Available from: http://www.ncbi.nlm.nih.gov/pubmed/24749759

[132] Devos D, Lebouvier T, Lardeux B, Biraud M, Rouaud T, Pouclet H, et al. Colonic inflammation in Parkinson's disease. Neurobiology of Disease [Internet]. 2013;**50**:42-48. Available from: http://www.ncbi.nlm.nih.gov/pubmed/23017648

[133] Fang X, Xu RS. Protective effect of simvastatin on impaired intestine tight junction protein ZO-1 in a mouse model of Parkinson's disease. Journal of Huazhong University of Science and Technology. Medical Sciences

[Internet]. 2015;**35**(6):880-884. Available from: http://www.ncbi.nlm.nih.gov/entrez/query.fcgi?cmd=Retrieve&db=PubMed&dopt=Citation&list_uids=26670440

[134] Schwiertz A, Spiegel J, Dillmann U, Grundmann D, Burmann J, Fassbender K, et al. Fecal markers of intestinal inflammation and intestinal permeability are elevated in Parkinson's disease. Parkinsonism & Related Disorders [Internet]. 2018;**50**:104-107. Available from: http://www.ncbi.nlm.nih.gov/entrez/query.fcgi?cmd=Retrieve&db=PubMed&dopt=Citation&list_uids=29454662

[135] Adams-Carr KL, Bestwick JP, Shribman S, Lees A, Schrag A, Noyce AJ. Constipation preceding Parkinson's disease: A systematic review and meta-analysis. Journal of Neurology, Neurosurgery, and Psychiatry [Internet]. 2016;**87**(7):710-716. Available from: http://www.ncbi.nlm.nih.gov/entrez/query.fcgi?cmd=Retrieve&db=PubMed&dopt=Citation&list_uids=26345189

[136] Chaudhuri KR, Odin P. The challenge of non-motor symptoms in Parkinson's disease. Progress in Brain Research [Internet]. 2010;**184**:325-341. Available from: http://www.ncbi.nlm.nih.gov/entrez/query.fcgi?cmd=Retrieve&db=PubMed&dopt=Citation&list_uids=20887883

[137] Fasano A, Visanji NP, Liu LW, Lang AE, Pfeiffer RF. Gastrointestinal dysfunction in Parkinson's disease. Lancet Neurology [Internet]. 2015;**14**(6):625-639. Available from: http://www.ncbi.nlm.nih.gov/entrez/query.fcgi?cmd=Retrieve&db=PubMed&dopt=Citation&list_uids=25987282

[138] Berg D, Postuma RB, Adler CH, Bloem BR, Chan P, Dubois B, et al. MDS research criteria for prodromal Parkinson's disease. Movement Disorders [Internet]. 2015;**30**(12):1600-1611. Available from: http://www.ncbi.nlm.nih.gov/entrez/query.fcgi?cmd=Retrieve&db=PubMed&dopt=Citation&list_uids=26474317

[139] Seguella L, Capuano R, Sarnelli G, Esposito G. Play in advance against neurodegeneration: Exploring enteric glial cells in gut-brain axis during neurodegenerative diseases. Expert Review of Clinical Pharmacology [Internet]. 2019;**12**(6):555-564. Available from: http://www.ncbi.nlm.nih.gov/entrez/query.fcgi?cmd=Retrieve&db=PubMed&dopt=Citation&list_uids=31025582

[140] Klingelhoefer L, Reichmann H. Pathogenesis of Parkinson disease–The gut-brain axis and environmental factors. Nature Reviews. Neurology [Internet]. 2015;**11**(11):625-636. Available from: http://www.ncbi.nlm.nih.gov/entrez/query.fcgi?cmd=Retrieve&db=PubMed&dopt=Citation&list_uids=26503923

[141] Lane CA, Hardy J, Schott JM. Alzheimer's disease. European Journal of Neurology [Internet]. 2018;**25**(1):59-70. Available from: http://www.ncbi.nlm.nih.gov/entrez/query.fcgi?cmd=Retrieve&db=PubMed&dopt=Citation&list_uids=28872215

[142] Leblhuber F, Geisler S, Steiner K, Fuchs D, Schutz B. Elevated fecal calprotectin in patients with Alzheimer's dementia indicates leaky gut. Journal of Neural Transmission (Vienna) [Internet]. 2015;**122**(9):1319-1322. Available from: http://www.ncbi.nlm.nih.gov/entrez/query.fcgi?cmd=Retrieve&db=PubMed&dopt=Citation&list_uids=25680441

[143] Joachim CL, Mori H, Selkoe DJ. Amyloid beta-protein deposition in tissues other than brain in Alzheimer's disease. Nature [Internet]. 1989; **341**(6239):226-230. Available from:

http://www.ncbi.nlm.nih.gov/entrez/query.fcgi?cmd=Retrieve&db=PubMed&dopt=Citation&list_uids=2528696

[144] Arai H, Lee VM, Messinger ML, Greenberg BD, Lowery DE, Trojanowski JQ. Expression patterns of beta-amyloid precursor protein (beta-APP) in neural and nonneural human tissues from Alzheimer's disease and control subjects. Annals of Neurology [Internet]. 1991;**30**(5):686-693. Available from: http://www.ncbi.nlm.nih.gov/entrez/query.fcgi?cmd=Retrieve&db=PubMed&dopt=Citation&list_uids=1763893

[145] Holmes C, Cunningham C, Zotova E, Woolford J, Dean C, Kerr S, et al. Systemic inflammation and disease progression in Alzheimer disease. Neurology [Internet]. 2009;**73**(10):768-774. Available from: http://www.ncbi.nlm.nih.gov/entrez/query.fcgi?cmd=Retrieve&db=PubMed&dopt=Citation&list_uids=19738171

[146] Solleiro-Villavicencio H, Rivas-Arancibia S. Effect of chronic oxidative stress on Neuroinflammatory response mediated by CD4(+)T cells in neurodegenerative diseases. Frontiers in Cellular Neuroscience [Internet]. 2018;**12**:114. Available from: http://www.ncbi.nlm.nih.gov/entrez/query.fcgi?cmd=Retrieve&db=PubMed&dopt=Citation&list_uids=29755324

[147] da Silveira AB, de Oliveira EC, Neto SG, Luquetti AO, Fujiwara RT, Oliveira RC, et al. Enteroglial cells act as antigen-presenting cells in chagasic megacolon. Human Pathology [Internet]. 2011;**42**(4):522-532. Available from: http://www.ncbi.nlm.nih.gov/entrez/query.fcgi?cmd=Retrieve&db=PubMed&dopt=Citation&list_uids=21208643

[148] Geboes K, Rutgeerts P, Ectors N, Mebis J, Penninckx F, Vantrappen G, et al. Major histocompatibility class II expression on the small intestinal nervous system in Crohn's disease. Gastroenterology [Internet]. 1992;**103**(2):439-447. Available from: http://www.ncbi.nlm.nih.gov/pubmed/1378803

[149] Turco F, Sarnelli G, Cirillo C, Palumbo I, De Giorgi F, D'Alessandro A, et al. Enteroglial-derived S100B protein integrates bacteria-induced toll-like receptor signalling in human enteric glial cells. Gut [Internet]. 2014;**63**(1):105-115. Available from: http://www.ncbi.nlm.nih.gov/pubmed/23292665

[150] Gabanyi I, Muller PA, Feighery L, Oliveira TY, Costa-Pinto FA, Mucida D. Neuro-immune interactions drive tissue programming in intestinal macrophages. Cell [Internet]. 2016;**164**(3):378-391. Available from: http://www.ncbi.nlm.nih.gov/pubmed/26777404

[151] De Schepper S, Verheijden S, Aguilera-Lizarraga J, Viola MF, Boesmans W, Stakenborg N, et al. Self-maintaining gut macrophages are essential for intestinal homeostasis. Cell. 2018;**175**(2):400-415

[152] Cardoso V, Chesne J, Ribeiro H, Garcia-Cassani B, Carvalho T, Bouchery T, et al. Neuronal regulation of type 2 innate lymphoid cells via neuromedin U. Nature. 2017 Sep;**549**(7671):277-281

[153] Klose CSN, Mahlakoiv T, Moeller JB, Rankin LC, Flamar A-L, Kabata H, et al. The neuropeptide neuromedin U stimulates innate lymphoid cells and type 2 inflammation. Nature. 2017 Sep;**549**(7671):282-286

[154] Ibiza S, Garcia-Cassani B, Ribeiro H, Carvalho T, Almeida L, Marques R, et al. Glial-cell-derived neuroregulators control type 3 innate lymphoid cells and gut defence. Nature [Internet]. 2016;**535**(7612):440-443.

Available from: http://www.ncbi.nlm.nih.gov/pubmed/27409807

[155] Veiga-Fernandes H, Coles MC, Foster KE, Patel A, Williams A, Natarajan D, et al. Tyrosine kinase receptor RET is a key regulator of Peyer's patch organogenesis. Nature. 2007;**446**(7135):547-551

[156] Capoccia E, Cirillo C, Gigli S, Pesce M, D'Alessandro A, Cuomo R, et al. Enteric glia: A new player in inflammatory bowel diseases. International Journal of Immunopathology and Pharmacology [Internet]. 2015;**28**(4):443-451. Available from: http://www.ncbi.nlm.nih.gov/pubmed/26526203

[157] Kermarrec L, Durand T, Gonzales J, Pabois J, Hulin P, Neunlist M, et al. Rat enteric glial cells express novel isoforms of Interleukine-7 regulated during inflammation. Neurogastroenterology and Motility: The Official Journal of the European Gastrointestinal Motility Society. 2019;**31**(1):e13467

[158] Kermarrec L, Durand T, Neunlist M, Naveilhan P, Neveu I. Enteric glial cells have specific immunosuppressive properties. Journal of Neuroimmunology [Internet]. 2016;**295-296**:79-83. Available from: http://www.ncbi.nlm.nih.gov/pubmed/27235353

[159] Koretz K, Momburg F, Otto HF, Moller P. Sequential induction of MHC antigens on autochthonous cells of ileum affected by Crohn's disease. The American Journal of Pathology [Internet]. 1987;**129**(3):493-502. Available from: http://www.ncbi.nlm.nih.gov/pubmed/3425689

[160] Ruhl A, Franzke S, Collins SM, Stremmel W. Interleukin-6 expression and regulation in rat enteric glial cells. American Journal of Physiology. Gastrointestinal and Liver Physiology [Internet]. 2001;**280**(6):G1163-G1171. Available from: http://www.ncbi.nlm.nih.gov/pubmed/11352809

[161] Ruhl A, Franzke S, Stremmel W. IL-1beta and IL-10 have dual effects on enteric glial cell proliferation. Neurogastroenterology and Motility: The Official Journal of the European Gastrointestinal Motility Society. [Internet]. 2001 Feb;**13**(1):89-94

[162] Rao M, Rastelli D, Dong L, Chiu S, Setlik W, Gershon MD, et al. Enteric glia regulate gastrointestinal motility but are not required for maintenance of the epithelium in mice. Gastroenterology [Internet]. 2017;**153**(4):1068-1081. Available from: http://www.ncbi.nlm.nih.gov/pubmed/28711628

[163] Aube AC, Cabarrocas J, Bauer J, Philippe D, Aubert P, Doulay F, et al. Changes in enteric neurone phenotype and intestinal functions in a transgenic mouse model of enteric glia disruption. Gut [Internet]. 2006;**55**(5):630-637. Available from: http://www.ncbi.nlm.nih.gov/pubmed/16236773

[164] Nasser Y, Fernandez E, Keenan CM, Ho W, Oland LD, Tibbles LA, et al. Role of enteric glia in intestinal physiology: Effects of the gliotoxin fluorocitrate on motor and secretory function. American Journal of Physiology. Gastrointestinal and Liver Physiology [Internet]. 2006;**291**(5):G912-G927. Available from: http://www.ncbi.nlm.nih.gov/pubmed/16798727

Chapter 4

Involvement of Astrocytes in the Process of Metabolic Syndrome

Ana Karolina Santana Nunes, Haniel Soares Fernandes and Sura Wanessa Santos Rocha

Abstract

Astrocytes constitute a very heterogeneous population of cells, which regulate pH, extracellular levels of ions and neurotransmitters, and energy metabolism in addition to actively participating in neurotransmission. In situations of damage to the CNS, the typical response is the degree of reactive gliosis, which can form glial scars. On the other hand, chronic diseases such as obesity, type 2 diabetes, hypertension, and atherosclerosis have been causally related to low-grade chronic inflammation in various metabolic tissues. It has been pointed out that the identification of hypothalamic inflammatory alterations are triggered by overnutrition, orchestrated by the hypothalamic immune system, and sustained by the pathophysiology associated with the metabolic syndrome. We discuss here the effects of astrocytes and the main astrocyte mechanisms involved in the metabolic syndrome and its comorbidities.

Keywords: astrocytes, neuroinflammation, metabolic syndrome, metabolism

1. Introduction

The cases of metabolic syndrome (MetS) in adults are increasing, due to several factors such as aging population, physical inactivity, obesity and chronic overnutrition [1]. Metabolic abnormalities are involved in the metabolic syndrome, such as diabetes mellitus, hypercholesterolemia and dyslipidemia, hypertension, and central obesity [2]. Several studies in neuroscience and immunology are linked to overnutrition to neuroinflammation, particularly in the hypothalamus and in the hippocampus, due to interaction between accelerated adiposity, hyperglycemia, and cognitive decline [3–5].

The primary risk of cognitive decline in both obese and hyperglycemic individuals is the systemic and chronic inflammatory component of MetS, due to preparation of the resident population of glial cells to establish a form of low-grade neuroinflammation [6].

Astrocytes are active agents of the dynamic central nervous system (CNS) signaling. The astrocytes participate in a variety of essential physiological processes in the healthy brain, such as providing structural support to neurons, participating in the formation and maturation of synapses, control of homeostasis of ions and metabolites, receptor trafficking, neurotransmitter clearance, and modulating the synaptic plasticity moment by moment [7]. Many studies have shown their contribution to information processing and memory formation in the brain, thus pointing to a role for astrocytes in higher integrated brain functions [8, 9].

A vast arsenal at the disposal of astrocytes is being defined, as in the determination of functions and mechanisms of reactive astrogliosis, cellular hypertrophy, and glial scar formation with preservation of the cellular domains and rearrangement of the tissue structure as well as contributing to specific CNS disorders and lesions [9]. Neuroinflammation in the enteric system occurs due to the activation of enteric glial cells (EGCs) which are the most abundant cells within the enteric nervous system (ENS). EGCs are located adjacent to the neurons within the enteric ganglia and along the interganglionic connections of the myenteric and submucosal plexus but also protrude into the extraganglionic mucosal layer [10–13]. Their morphology and the expression of markers such as calcium-binding protein S100 and glial fibrillary acidic protein (GFAP) [14] resemble central nervous system astrocytes.

The EGCs can exert immunomodulatory functions; they can secrete inflammatory signaling molecules such as interleukins IL-1β and IL-6 [10, 15] as well as other mediators, including nerve growth factor (NGF), S-nitrosoglutathione (GSNO), nitric oxide (NO), and S100B [16], and express class II major histocompatibility (MHC) complex molecules [16, 17]. CNS astrocytes activated by inflammation are characterized by hypertrophy and proliferation, coupled with a positive regulation of the GFAP's cytoskeleton [18], due metabolic syndrome and inflammatory conditions of the intestinal disease (IBD) [19].

Many studies have shown that multiple connections between peripheral and cerebral changes involving inflammatory, metabolic, and neural components have been identified under conditions associated with obesity [20–22]. Therefore, identifying and treating these conditions is of primary importance to people worldwide. Obesity-related chronic inflammation provides an important link to metabolic derangements including insulin resistance, cognitive impairment affecting the hypothalamus, and other brain regions [20].

The present study was designed to discuss the effects of astrocytes and the main astrocyte and neuroinflammatory mechanisms involved in the metabolic syndrome and their comorbidities, which gave rise to the field of immunometabolism.

2. Neuroimmunologic mechanisms and role of glial cells

Until recently the CNS was considered to be immunologically privileged, since many antibodies and peripheral immune system cells are usually blocked through the blood-brain barrier (BBB), a specialized structure composed of endothelial cells (ECs), pericytes, astrocytes, and microglia [23]. The BBB maintains the chemical composition of the neuronal microenvironment, which is necessary for the proper functioning of neuronal circuits, synaptic transmission, synaptic remodeling, angiogenesis, and neurogenesis [24].

The immune system influences the functioning of BBB, which in turn affects the functioning of the CNS in both physiological and pathological conditions. In some cases, the BBB separates the CNS from the immune system; in others it acts as a mediator of neuroimmune interactions, and in others it may act as a target for immune system attacks [25]. In physiological conditions, immune cells cross the BBB at a very low rate, through specific interactions, promoting the endothelial junctions that control the flow of cells through them [26–28]. On the other hand, neuropathological diseases, such as Alzheimer's disease (AD), Parkinson's disease (PD), lateral amyotrophic sclerosis (LAS), multiple sclerosis (MS), and BBB destruction or damage, can be induced or mediated by LPS (lipopolysaccharide), cytokines, prostaglandins, and nitric oxide. The BBB is capable of responding to LPS due to the presence of Toll-like receptor 4 (TLR4) and other Toll-like receptors on the membranes of the BBB cells [29]. Similarly, these cells have receptors for

DOI: http://dx.doi.org/10.5772/intechopen.87931

cytokines, chemokines, and other immunological molecules [30, 31]. As a result, the immune system is able to affect the functions of the BBB beyond those of disruption.

Perivascular cells (astrocytes and microglia) in addition to endothelial cells produce several inflammatory factors, such as the release of cytokines and chemokines that affect the BBB permeability and the expression of adhesion molecules. Cytokines (TNF-α, IL1-β, IFN-γ) can stimulate the expression of adhesion molecules (vascular cell adhesion molecule 1 (VCAM-1) and intercellular adhesion molecule 1(ICAM-1)) in endothelial cells allowing the passage of activated leukocytes into the CNS [32]. Immune cell (macrophages, lymphocytes) traffic through the BBB should initiate or contribute to a vicious cycle resulting in progressive synaptic dysfunction and neuronal loss in neurodegenerative disorders [20, 24].

On the other hand, a specific subset of T cells is essential to suppress autoimmunity and maintain immune homeostasis. T regulatory cells (Treg) have been characterized with important functions. Emerging evidence shows that Treg cells are not only important for maintaining immune balance at the periphery but also contribute to the self-tolerance and immune privilege in CNS [33]. Leukocyte extravasation requires interactions between adhesion molecules in endothelial cells and leukocytes. Leukocyte adhesion molecules (LAMs) expressed by ECs include P-selectin, E-selectin, intercellular adhesion molecule 1 (ICAM-1), and vascular cell adhesion molecule 1 (VCAM-1) [34]. Selectin binds to P-selectin-binding glycoprotein (PSGL-1), while ICAM-1 and VCAM-1 bind to α4-integrins in leukocytes. After the initial binding event, immune cells roll along the vessel wall releasing chemokines that strengthen their binding interactions, promoting the state of neuroinflammation [35].

Neuroinflammation is recognized as a prominent feature of various pathological conditions [36]. Thus, several lines of evidence strongly suggest that neuroinflammation is a crucial process involved in the progression of neuronal degeneration, a common feature observed in several neurodegenerative disorders such as degenerative neuropathologies [36]. In the inflammatory process, the main cellular events are observed, such as increased blood flow and vascular permeability with consequent venular dilation and recruitment of cells to the inflamed site. A significant role played by reactive oxygen species has been observed to develop inflammation, causing endothelial cell damage and increased microvascular permeability, chemotactic factor production, neutrophil recruitment, oxidation, and lipid peroxidation [37]. Such inflammatory mediators play a regulatory role in the growth, differentiation, and activation of immune cells [38]. Glial cells (microglia, astrocytes, and oligodendrocytes) define cerebral homeostasis and are responsible for defense and preservation against neural tissue injury [14].

3. The role of astrocytes in neuroinflammation

Astrocytes are an important group of heterogeneous cells and play key roles in the physiology of the nervous system, including regulation of pH, extracellular levels of ions and neurotransmitters, and energy metabolism. In addition, it plays an important role in the formation and functioning of the BBB [39] and also actively participates in neurotransmission [27]. In pathological situations in the CNS, the typical response of astrocytes is the state of reactive gliosis involving positive gene regulation of cytoskeletal proteins (e.g., glial fibrillary acidic protein (GFAP)) and corresponding to the change in morphology reaching a state of hypertrophy, hyperplasia, and glial scar formation [40–42].

In addition, astrocytes play an important role in central immunity. The innate immune response is accurately adjusted by identifying the type of threat that is present. The molecular structures that are associated with threats are recognized by Pattern Recognition Receptors (PRRs). PRRs recognize molecular patterns associated with pathogens (PAMPs) are expressions such as microorganisms such as bacteria, viruses and viruses, and damage-associated diseases (DAMPs) that signal cellular damage and are therefore responsible for a state of stress or injury [14]. Among PRRs, one of the main classes is the family of transmembrane proteins of Toll-like receptors (TLRs). Generally pathogen response and tissue damage happen quickly, assuming some roles of the cells of the immune system, releasing cytokines (IL-1β, IL-6, TNF-α) and chemokines (MCP-1, CCR-2, COX-2), influencing other cells of the immune system (macrophages and lymphocytes), and modulating the BBB [14].

Among glial cells, astrocytes play a role in the release of Toll-like receptors [43]. Since the TLRs are expressed and detected by the binding of their binding genes, a signaling mediated by the myeloid differentiation gene 88 (Myd88) is initiated, having an activation of the nuclear transcription factor NFκB. In the activation of NFκB, the inflammatory process is released through the secretion of pro-inflammatory molecules (1β, IL-6, TNF-α, IL-12). In addition, there is no state of activation of astrocytes to recruit microorganisms, lymphocytes, and dendritic cells to the lesion site [44].

4. Implications of metabolic syndrome in the central nervous system: the role of the bowel-brain axis

The increased obesity in the last 40 years is considered a consequence of the sedentary lifestyle and adherence to diets rich in saturated fat and refined carbohydrates that induce changes in the microbiota and underlying metabolic and psychological complications [45]. Bacteria, viruses, protozoa, archaea, and fungi represent the microorganisms that inhabit the intestinal tract of mammals, with bacteria composing the majority [46, 47] in concentrations between 10^1 to 10^3 cells per gram in the upper intestine and 10^{11} to 10^{12} cells per gram in the colon [48].

These microorganisms play a role in human physiology through various mechanisms, such as the metabolism of nutrients and the regulation of immunological and neuroendocrine functions, as they bind to the CNS through the enteric nervous system (ENS) [49]. From the microbiota, the active metabolite LPS (lipopolysaccharide), in addition to short-chain fatty acids (SCFA), from invasive and commensal bacteria, respectively [50, 51], can be expressed in the intestinal lumen and influence the integrity of BBB [52], especially butyrate, in which it positively induces the expression of junction proteins, including claudin-2, occludin, cingulin, and occludens-1 and occludens-2 (ZO-1 and ZO-2), forming protein structures in the gut, like tight junction [53, 54]. These proteins form a mechanical link between epithelial cells and establish paracellular diffusion of fluids and solutes in the barrier [55]. Where high fiber and fruit meals were shown to reduce the increasesinduced by meals with high saturated fat and high carbohydrate content inlevels inflammatory response [56].

In mammals, colon epithelial cells, adipocytes, and peripheral blood mononuclear cells express a pair of G-protein-coupled receptors (GPR41 and GPR43) that are activated by SCFAs through the receptors on the T enteroendocrine cells [57]. They sense the amount of AGCC produced by bacteria in the colon and secrete the glucagon-like peptide 1 (GLP-1) and the tyrosine tyrosine peptide (PYY), allowing inhibition of intestinal motility and increased absorption of nutrients, respectively [58]. Thus, the composition of the diet determines the type of nutrient that reaches the gastrointestinal tract (GI) that can alter the composition of the intestinal microbiota and the production of metabolites to, consequently, influence intestinal permeability [59].

Bacterial products, such as LPS, can display a variety of PAMPs recognized by TLRs and nucleotide-binding oligomerization domain receptors (NOD) on macrophages and dendritic cells in the innate immune system, such as flagellin, recognized by TLR5 to induce α-defensin secretion through Paneth cells, a NOD-dependent antimicrobial protein [60]. All TLRs recognize protein, lipid, or nucleotide PAMPs. TLR2, TLR4, and TLR6 recognize fungal PAMPs, while TLR9 and TLR11 recognize protozoan PAMPs [61]. Although TLRs can activate immune cell proliferation through an Akt-dependent pathway, they will all induce the expression and secretion of cytokines [61]. Although TLRs may activate immune cell proliferation via an Akt-dependent pathway, all but TLR3 will induce the expression and secretion of tumor necrosis factor-α (TNF-α), interleukin 1β (IL-1β), and interferon-γ (IFN-γ), recruiting the primary myeloid differentiation response 88 (MYD88), which recruits the kinase family associated with the interleukin-1 receptor (IRAK), leading to phosphorylation of the inhibitory protein kappa B (I- kB) to induce translocation of nuclear factor kappa B (NF-kB) and influence the expression of inflammation [62, 63]. In addition, the activation of TLRs participates in the proliferation of epithelial cells and IgA secretion in the intestinal lumen, essential for intestinal barrier integrity and bacterial population balance, respectively [61]. Deregulation of these processes, or excessive activation of TLRs, can result in chronic inflammatory responses and exuberant repair [64].

In the epithelium, segmented bacterial filaments (SFB) and other commensal microbes activate dendritic cells (DCs) and macrophages in the lamina propria, inducing T helper 17 (TH17) cells through the production of interleukins IL-1β, IL-6, and IL-23 and helper T cells (TH1) by the possible production of interleukin-12 (IL-12) [21]. TH17 cells regulate the gut microbiota community by secreting the IL-22-dependent antimicrobial lectin regenerating islet-derived 3 gamma (REGIIIγ). One of the microbial derivatives, polysaccharide A (PSA), stimulates intestinal epithelial cells to secrete growth factor and β transformation (TGFβ), inducing DCs and macrophages to secrete retinoic acid and interleukin 10 (IL-10) to promote activation of regulatory T cells and forkhead box P3 (FOXP3) and, subsequently, inactivation of TH17 and TH1 cells, in a type of negative feedback between the cells of defense in relation to the balance of the gut bacteria [65]. Activation of TLRs induces B cell-activating factor (BAFF) secretion, which differentiates B cells by increasing activation-induced cytidine deaminase (AID) expression and promotes differentiation of IgA-producing plasma cells, by maturing antibodies and casting in the intestinal lumen to alter the composition and function of the microbiota [65].

In obese individuals, due to imbalance of the microbiota (dysbiosis) and oxidative stress, TLR4 is activated and recognizes the bacterial LPS and the flagellin of commensal bacteria, activating TLR5 for dendritic cell signaling and activation of innate lymphoid cells (ILCs), both processes to secrete REGIIIY [66]. ILCs communicate with the microbiota through cytokines, aryl hydrocarbon (AhR) receptors, and antimicrobial peptides and participate in the cross-talk of epithelial cells with the intestinal microbiota [67]. Divided into three groups, group ILCs (ILC1) are activated by interleukin-12 (IL-12) derived from myeloid cells, which in response secrete IFN-γ, whereas group ILCs (ILC2) interact with mast cells, eosinophils, basophils, and macrophages, and group 3 (ILC3) ILCs interact with cells of both innate and adaptive immune systems to secrete IL-22 and initiate an antimicrobial program along with restoration of the intestinal barrier [68].

5. Neuroinflammation from the intestine

The gut-brain axis is composed of the central nervous system, innervated by afferents and efferents of the vagus nerve and extrinsic fibers of the autonomic nervous system (ANS) interconnecting it to the hypothalamic-pituitary-adrenal (HPA) axis, in addition to the intrinsic neurons of the enteric nervous system and the intestinal microbiota. Intrinsic intestinal innervations connect the intestine to the brain via vagal and spinal fibers, whereas the brain sends efferent sympathetic and parasympathetic fibers to the intestine [69, 70]. The HPA axis is part of the limbic system and is the main regulator of stress response and intestinal function during digestion which, due to the corticotrophin releasing factor (CRF) secreted by the hypothalamus, can influence the motility, permeability, and level of intestinal inflammation [47, 71]. Therefore stress and emotions can influence the microbial composition of the intestine by bacterial products that gain access through the bloodstream and the postrema area and due to the release of glucocorticoids and/or sympathetic neurotransmitters that influence the physiology of the intestine and alter the habitat of the microbiota, such as noradrenaline, which may even influence bacterial gene expression or signaling among bacteria, altering the composition and activity of the microbiota [72].

After bacterial colonization, increased production of neurotransmitters, such as serotonin (5-HT) and γ-aminobutyric acid (GABA), and the expression of various cytokines are physiological implications essential for intestinal homeostasis and HPA axis programming, which plays an important role in stress responses [73]. The attention is focused on the stress due to serotonin, which is synthesized through tryptophan, in enterochromaffin cells (EC), about 90%, and in autonomic nerves, about 10%, i.e., at the level of the gastrointestinal tract [74].

Stress, corticosterone, and inflammation are the cornerstones of the catabolism of L-tryptophan (TRP) to kynurenine (KYN) and, subsequently, to quinolinic acid (QUIN) [75, 76]. TRP catalysis occurs through the enzymes indoleamine 2,3-dioxygenase (IDO), kynurenine monooxygenase (KMO), and tryptophan 2,3-dioxygenase (TDO), where stress results in the production of corticotropin-release hormone (CRH) by the hypothalamus, which induces the synthesis of adrenocorticotropic hormone (ACTH) by corticotrope cells in the anterior pituitary to target by blood the adrenal cortex and synthesize glucocorticoids for induction of TDO and activation of intracellular glucocorticoid receptors (GR) and subsequent TRP catalysis in KYN and kynurenic acid (KYNA) by kynurenine aminotransferase (KAT) or in 3-hydroxyquinurenine (3OH-KYN), both with neurotoxic potential of catabolizing 5-HT in 5-hydroxyindoleacetic acid (5-HIAA) by kynurenine monooxygenase [77]. On the other hand, stress induced by β-adrenergic receptors on the MSA axis (medullary sympathetic- adrenal) activates lymphoid cells and induces the release of proinflammatory cytokines IL-1β, IL-6 and IFN-γ and catecholamines [76].

These events induce the barrier permeability and increase of bacterial endotoxin through the gut, which stimulates immune cells in the lamina propria to secrete proinflammatory cytokines and prostaglandins (PGE2) to communicate the brain via afferent nerves, compromising the intestinal barrier and creating a cycle, where inflammatory cytokines will activate the SAM and HPA axes, resulting in barrier rupture, increased endotoxin translocation, and an inflammatory and stress state [78]. Where, one of the causes of this translocation, from the intestinal point of view, concerns LPS of invasive bacteria and their arrival in the intestinal lumen, in which it will induce a pro-inflammatory response in lymphoid and innate immune cells, and subsequent release of cytokines such such as IL-1β, IL-6 and TNF-α, in the bloodstream to the brain, from the axis of the gut-brain, affecting BBB integrity, causing a pro-inflammatory stress cycle in CNS cells inducing neuroinflammation [79] **Figure 1**.

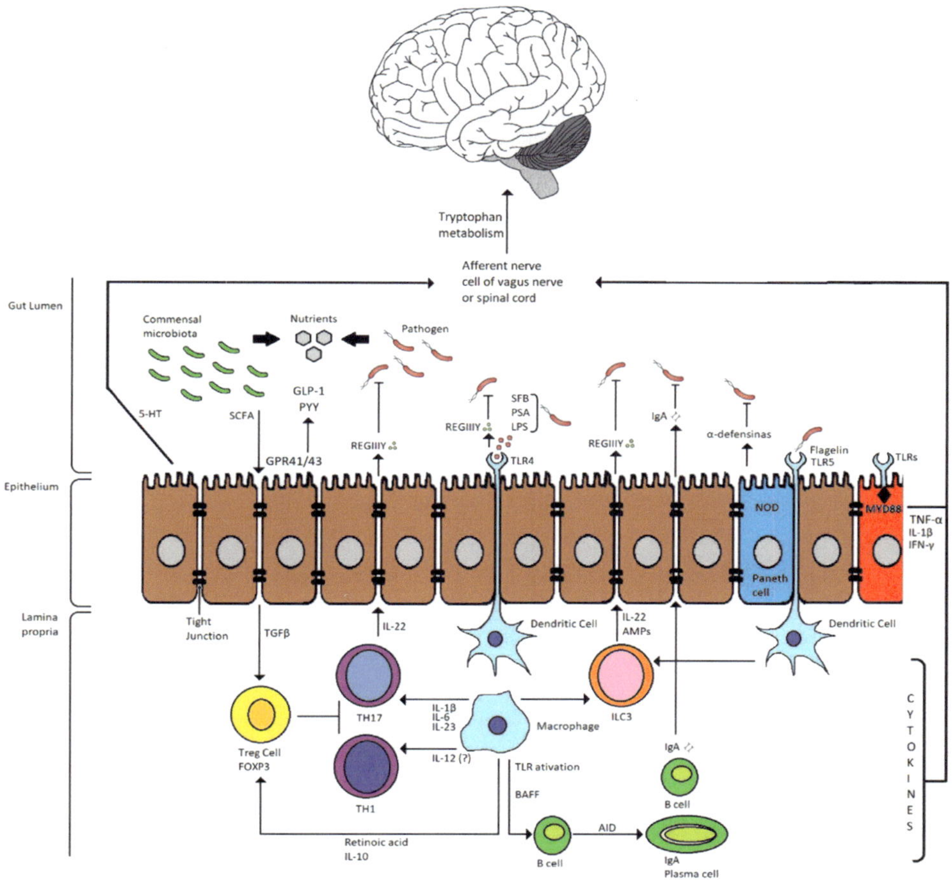

Figure 1.
Mechanisms by which the intestine is able to influence actions in the central nervous system (Fernandes, HS and Nunes, AKS).

6. Mechanisms of neuroinflammation and the process of metabolic syndrome

The proinflammatory cytokines (IL-1β, IL-6 and IFN-γ) and catecholamines, are able to induce IDO within CNS, exactly within the astrocyte and microglia, so the uptake and metabolism of L-tryptophan into theses cells leads to the production of KYNA, was report have neuroprotective actions in the CNS, but the catabolism of L-tryptophan in the microglia gives rise to metabolites with reactive oxidative properties, such as 3OH-KYN, 3-HAA and QUIN, that can be transported through the BBB to serve as substrates and contribute to the kinurenine pathway in the CNS, where macrophages and microglia represent the main sources of QUIN, a agonist on N glutamate receptor subtype methyl-D-aspartate (NMDA) that acts as a being able contribute to excitotoxicity and neurotoxicity [80, 81].

The astrocytes do not appear to have KMO, which favors the formation of KYNA, which after being released in the presynaptic area preferentially inhibits the NMDARs and α7 nicotinic acetylcholine receptors (nAChRs) in the extra-synaptic [81] **Figure 2**.

This leads to more leukocyte inflammation and infiltration at the BBB, since tight junction proteins, including claudins, ZO-1, and occludins, are downregulated [82].

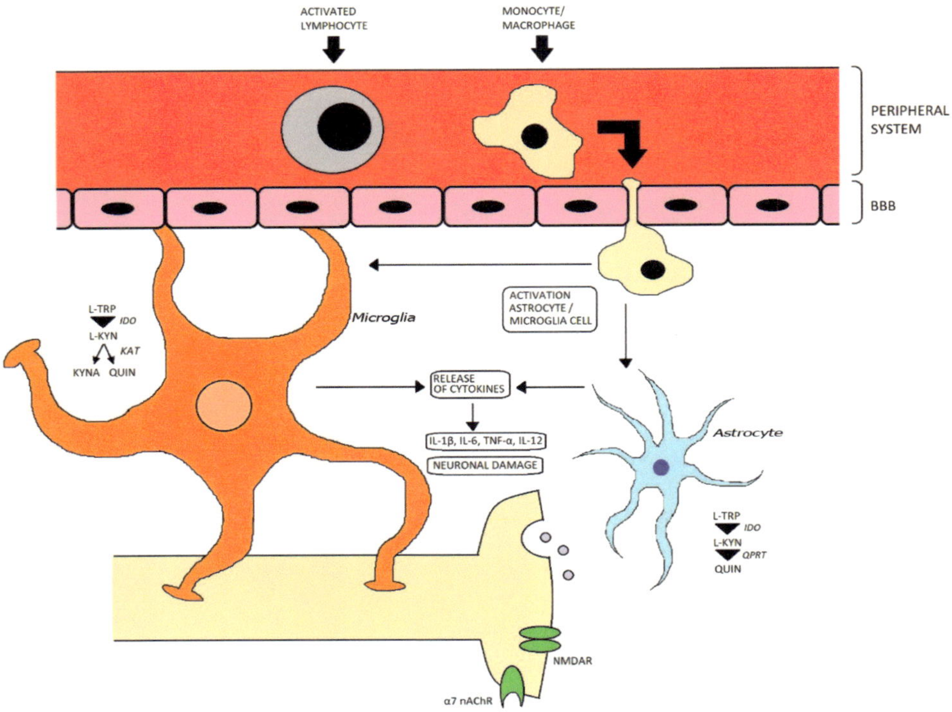

Figure 2.
Mechanism of astrocytes and microglia during the metabolic syndrome. Increased BBB allows the invading cells of the peripheral immune system and promotes the activation of glial cells, with consequent release of cytokines. In parallax, description of the KYN mechanism (Fernandes, HS and Nunes, AKS).

In response to inflammation, leukocyte extravasation increases with positive regulation of VCAM-1 and ICAM-1 [39]. Once the functional capacity of astrocytes is compromised, BBB is impaired, resulting in a significant increase in BBB permeability rate, a promotion of leukocytes in cerebrospinal fluid (CSF), and increased immune response, including pathogens and toxins in the CNS. This process favors the activation of astrocytes and microglia, which stimulates the continuous symptoms in the CNS, including the hypothalamus, with consequent response to the stress of insulin symptoms, as well as cognitive injury [83].

Within the metabolic syndrome, diabetes mellitus involves the CNS, and insulin signal transduction involves the activation of phosphatidyl-3 kinase (PI3K) and mitogen-activated protein kinase (MAPK) pathways [84]. The insulin receptor (IR) has two subunits (α and β). The α-subunit is directed toward the extracellular medium. Insulin binding to the IR receptor promotes autophosphorylation of the receptor on β subunits, located within the plasma membrane. A phosphorylation cascade of the insulin receptor substrates (IRS) 1 to 4 is then followed, and PI3K is then recruited into the membrane and induces the insertion of the GLUT glucose transporter into the plasma membrane [85]. In addition, PI3K also phosphorylates AKT, which in turn phosphorylates the glycogen synthase kinase 3 (GSK3) protein by inactivating it. This inactivation decreases the phosphorylation of Tau, which is present in neurons and has the function of stabilizing the microtubules for the transport of synaptic vesicles and other cellular components [86]. Therefore, the neuroinflammatory process comprises several mechanisms through the activation of the glial cells that leads to neuronal damage and consequent damages to the CNS.

7. Conclusion

In metabolic syndrome and obesity-associated conditions, immune and metabolic dysregulation results in chronic systemic inflammation, neuroinflammation, cognitive impairment, and other pathological manifestations. An understanding of this complex pathology requires providing new insight into the regulatory role of the astrocytes.

Although research to date in the fields of immunometabolism and neuroinflammation has produced encouraging preliminary results, there remains a vast expanse of unexplored questions requiring the interdisciplinary knowledge of metabolism, neuroscience, and immunology.

In summary, the descriptions of this study indicate that astrocytes play an important role in immunity by triggering neuroinflammation mediated by metabolic syndrome associated with obesity. Elucidating these mechanisms by binding the metabolism syndrome, inflammation, and CNS by astrocytes could generate potential new therapeutic targets or specific strategies to combat metabolic syndrome and obesity.

Abbreviation

ACTH	adrenocorticotropic hormone
ANS	autonomic nervous system
CNS	central nervous system
DAMPs	damage-associated diseases
EGCs	enteric glial cells
ENS	enteric nervous system
GFAP	glial fibrillary acidic protein
GI	gastrointestinal tract
GPR	G-protein-coupled receptors
GR	glucocorticoid receptors
GSNO	nitrosoglutathione
I- kB	inhibitory protein kappa B
IBD	intestinal disease
ICAM	intercellular adhesion molecules 1
ILCs	innate lymphoid cells
IRS	insulin receptor substrates
KMO	kynurenine monooxygenase
KYN	kynurenine
KYNA	kynurenic acid
MAPK	mitogen-activated protein kinase
MetS	metabolic syndrome
MHC	class II major histocompatibility
Myd88	myeloid differentiation gene 88
NF-kB	nuclear factor kappa B
NGF	nerve growth factor
NMDA	methyl-D-aspartate
NO	nitric oxide
NOD	oligomerization domain receptors
PAMPs	molecular patterns associated with pathogens
PRRs	pattern recognition receptors
QUIN	kynolinic acid
SCFA	short-chain fatty acids

SFB	segmented bacterial filaments
TLR4	Toll-like receptor 4
VCAM	vascular cell adhesion molecule 1

Author details

Ana Karolina Santana Nunes[1,2*], Haniel Soares Fernandes[3]
and Sura Wanessa Santos Rocha[2,4]

1 Federal University of Pernambuco, Recife, Pernambuco, Brazil

2 Ultrastructure Laboratory, Aggeu Magalhães Research Center (FIOCRUZ), Recife, Pernambuco, Brazil

3 Estacio de Sa University, Fortaleza, Ceará, Brazil

4 University of Pernambuco, Recife, Pernambuco, Brazil

*Address all correspondence to: nunes.aks@gmail.com

References

[1] James AM, Collins Y, Logan A, Murphy MP. Mitochondrial oxidative stress and the metabolic syndrome. Trends in Endocrinology and Metabolism. 2012;**23**(9):429-434

[2] Lavie CJ, Milani RV, Ventura HO. Obesity and cardiovascular disease. Risk factor, paradox, and impact of weight loss. Journal of the American College of Cardiology. 2009;**53**(21):1925-1932

[3] Cai D. Neuroinflammation and neurodegeneration in overnutrition-induced diseases. Trends in Endocrinology and Metabolism. 2013;**24**(1):40-47

[4] Moreno-Navarrete JM, Blasco G, Puig J, Biarnés C, Rivero M, Gich J, et al. Neuroinflammation in obesity: Circulating lipopolysaccharide-binding protein associates with brain structure and cognitive performance. International Journal of Obesity. 2017;**41**(11):1627-1635

[5] Wanrooy BJ, Kumar KP, Wen SW, Qin CX, Ritchie RH, Wong CHY. Distinct contributions of hyperglycemia and high-fat feeding in metabolic syndrome- induced neuroinflammation. Journal of Neuroinflammation. 2018;**8**:1-13

[6] Drake C, Boutin H, Jones MS, Denes A, McColl BW, Selvarajah JR, et al. Brain inflammation is induced by co-morbidities and risk factors for stroke. Brain, Behavior, and Immunity. 2011;**25**(6):1113-1122

[7] Unger JW. Glial reaction in aging and Alzheimer's disease. Frontiers in Aging Neuroscience. 1998;**28**:24-28

[8] Hatten ME, Liem RKH, Shelanski ML, Mason CA. Astroglia in CNS injury. Glia. 1991;**243**:233-243

[9] Sofroniew MV, Vinters HV. Astrocytes: Biology and pathology. Acta Neuropathologica. Jan 2010;**119**(1): 7-35. DOI: 10.1007/s00401-009-0619-8. [Epub 2009 Dec 10]

[10] Ruhl A. Glial cells in the gut. Neurogastroenterology and Motility. Acta Neuropathologica. 2005;**17**(6):777-790

[11] Opazo MC, Ortega-Rocha EM, Coronado-Arrázola I, Bonifaz LC, Boudin H, Neunlist M, et al. Intestinal microbiota influences non-intestinal related autoimmune diseases. Frontiers in Microbiology. 2018;**9**:1-20

[12] Bohórquez DV, Samsa LA, Roholt A, Medicetty S, Chandra R, Liddle RA. An enteroendocrine cell-enteric glia connection revealed by 3D electron microscopy. PLoS One. 26 Feb 2014;**9**(2):e89881. DOI: 10.1371/journal.pone.0089881.eCollection 2014

[13] Boesmans W, Lasrado R, Vanden Berghe P, Pachnis V. Heterogeneity and phenotypic plasticity of glial cells in the mammalian enteric nervous system. Glia. 2015;**63**(2):229-241

[14] Jessen K, Mirsky R. Astrocyte-like glia in the peripheral nervous system: An immunohistochemical study of enteric glia. The Journal of Neuroscience. 2018;**3**(11):2206-2218

[15] Murakami M, Ohta T, Ito S. Lipopolysaccharides enhance the action of bradykinin in enteric neurons via secretion of interleukin-1β from enteric glial cells. Journal of Neuroscience Research. 2009;**87**(9):2095-2104

[16] Cirillo C, Sarnelli G, Turco F, Mango A, Grosso M, Aprea G, et al. Proinflammatory stimuli activates human-derived enteroglial cells

and induces autocrine nitric oxide production. Neurogastroenterology and Motility. 2011;**23**(9):10-12

[17] Cirillo C, Sarnelli G, Esposito G, Grosso M, Petruzzelli R, Izzo P, et al. Increased mucosal nitric oxide production in ulcerative colitis is mediated in part by the enteroglial-derived S100B protein. Neurogastroenterology and Motility. Nov 2009;**21**(11):1209-e112. DOI: 10.1111/j.1365-2982.2009.01346.x. [Epub 2009 Jun 24]

[18] Reyes RC, Verkhratsky A, Parpura V. Plasmalemmal Na^{+}/Ca^{2+} exchanger modulates Ca^{2+} -dependent exocytotic release of glutamate from rat cortical astrocytes. ASN Neuro. 2012;**4**(1):AN20110059

[19] Chan KL, Cathomas F, Russo SJ. Central and peripheral inflammation link metabolic syndrome and major depressive disorder. Acta Neuropathologica. Physiology (Bethesda). 1 Mar 2019;**34**(2):123-133. DOI: 10.1152/physiol.00047.2018

[20] Guillemot-Legris O, Muccioli GG. Obesity-induced neuroinflammation: Beyond the hypothalamus. Trends in Neurosciences. 2017;**40**(4):237-253 . DOI: 10.1016/j.tins.2017.02.005. [Epub 2017 Mar 18]

[21] Schwartz MW, Porte D Jr. Diabetes, obesity, and the brain. Science. 2005;**307**:375-380

[22] Pavlov VA, Tracey KJ. The vagus nerve and the inflammatory reflex—Linking immunity and metabolism. Nature Reviews. Endocrinology. 2012;**8**(12):743-754

[23] Disabato D, Quan N, Godbout JP. Neuroinflammation: The devil is in the details. Journal of Neurochemistry. 2017;**139**(Suppl. 2):136-153

[24] Zlokovic BV. New therapeutic targets in the neurovascular pathway in Alzheimer's disease. Neurotherapeutics. 2008;**5**(3):409-414

[25] Erickson MA, Banks WA. Blood-brain barrier dysfunction as a cause and consequence of Alzheimer's disease. Journal of Cerebral Blood Flow and Metabolism. 2013;**33**(10):1500-1513

[26] Jagim AR, Harty PS, Camic CL. Common ingredient profiles of multi-ingredient pre-workout supplements. Nutrients. 2019;**11**(2):4-11

[27] Banks WA. The blood-brain barrier in neuroimmunology: Tales of separation and assimilation. Brain, Behavior, and Immunity. 2015;**44**:1-8

[28] Xu H, Manivannan A, Liversidge J, Sharp PF, Forrester JV, Crane IJ. Requirements for passage of T lymphocytes across non-inflamed retinal microvessels. Journal of Neuroimmunology. 2003;**142**(1-2):47-57

[29] Siednienko J, Halle A, Nagpal K, Golenbock DT, Miggin SM. TLR3-mediated IFN-β gene induction is negatively regulated by the TLR adaptor MyD88 adaptor-like. European Journal of Immunology. 2010;**40**(11):3150-3160

[30] Łabuzek K, Liber S, Gabryel B, Okopień B. Metformin has adenosine-monophosphate activated protein kinase (AMPK)-independent effects on LPS-stimulated rat primary microglial cultures. Pharmacological Reports. 2010;**62**(5):827-848

[31] Paintlia AS, Paintlia MK, Singh I, Singh AK. IL-4-induced peroxisome proliferator-activated receptor gamma activation inhibits NF-kappaB trans activation in central nervous system (CNS) glial cells and protects oligodendrocyte progenitors under neuroinflammatory disease conditions: Implication for CNS-demyelinating diseases. Journal of Immunology. 1 Apr 2006;**176**(7):4385-4398

[32] Hoermann R, Midgley JEM, Larisch R, Dietrich JW. Homeostatic control of the thyroid-pituitary axis: Perspectives for diagnosis and treatment. Frontiers in Endocrinology. 2015;**6**:1-17

[33] Amor S, Woodroofe MN. Innate and adaptive immune responses in neurodegeneration and repair. Immunology. 2014;**141**(3):287-291

[34] Lee KS, Kim J, Kwak SN, Lee KS, Lee DK, Ha KS, et al. Functional role of NF-κB in expression of human endothelial nitric oxide synthase. Biochemical and Biophysical Research Communications. 2014;**448**(1):101-107

[35] Gieske BT, Stecker RA, Smith CR, Witherbee KE, Harty PS, Wildman R, et al. Metabolic impact of protein feeding prior to moderate-intensity treadmill exercise in a fasted state: A pilot study. Journal of the International Society of Sports Nutrition. 2018;**15**(1):1-12

[36] Glass CK, Saijo K, Winner B, Marchetto MC, Gage FH. Mechanisms underlying inflammation in neurodegeneration. Cell. 2010;**140**(6):918-934

[37] Menegazzi M, Di Paola R, Mazzon E, Genovese T, Crisafulli C, Dal Bosco M, et al. Glycyrrhizin attenuates the development of carrageenan-induced lung injury in mice. Pharmacological Research. 2008;**58**(1):22-31

[38] Schins RPF, McAlinden A, MacNee W, Jimenez LA, Ross JA, Guy K, et al. Persistent depletion of I kappa B alpha and interleukin-8 expression in human pulmonary epithelial cells exposed to quartz particles. Toxicology and Applied Pharmacology. 2000;**167**(2):107-117

[39] Figley CR, Stroman PW. The role(s) of astrocytes and astrocyte activity in neurometabolism, neurovascular coupling, and the production of functional neuroimaging signals. The European Journal of Neuroscience. 2011;**33**(4):577-588

[40] Jessen R, Wood PM, Wanner INAB, Mahoney J, An K, Bates M, et al. Invariant mantling of growth cones by Schwann cell precursors characterize growing peripheral nerve fronts. Glia. 2006;**438**:424-438

[41] Rothermundt M, Peters M, Prehn JHM, Arolt V. S100B in brain damage and neurodegeneration. Microscopy Research and Technique. 2003;**60**(6):614-632

[42] Nagayach A, Patro N, Patro I. Astrocytic and microglial response in experimentally induced diabetic rat brain. Metabolic Brain Disease. 2014;**29**(3):747-761

[43] Owen JA, Kehoe SJ, Oliver SJ. Influence of fluid intake on soccer performance in a temperate environment. Journal of Sports Sciences. 2013;**31**(1):1-10

[44] González H, Elgueta D, Montoya A, Pacheco R. Neuroimmune regulation of microglial activity involved in neuroinflammation and neurodegenerative diseases. Journal of Neuroimmunology. 2014;**274**(1-2):1-13

[45] Portune KJ, Benítez-Páez A, Del Pulgar EM, Cerrudo V, Sanz Y. Gut microbiota, diet, and obesity-related disorders-The good, the bad, and the future challenges. Molecular Nutrition & Food Research. Jan 2017;**61**(1). DOI: 10.1002/mnfr.201600252. [Epub 2016 Sep 20]

[46] Xu J, Mahowald MA, Ley RE, Lozupone CA, Hamady M, Martens EC, et al. Evolution of symbiotic bacteria in the distal human intestine. PLoS Biology. 2007;**5**(7):1574-1586

[47] Gill SR, Pop M, Deboy RT, Eckburg PB, Peter J, Samuel BS, et al. Metagenomic analysis of the human

distal gut microbiome Steven. Science. 2011;**312**(5778):1355-1359

[48] Derrien M, van Hylckama Vlieg JET. Fate, activity, and impact of ingested bacteria within the human gut microbiota. Trends in Microbiology. 2015;**23**(6):354-366

[49] Backhed F, Ley RE, Sonnenburg JL, Peterson DA, Gordon JI. Host-bacterial mutualism in the human intestine - supplemental materials. Science. 2005;**307**(5717):1915-1920

[50] Bernat O. Medicines from microbiota. Nature Biotechnology. 2013;**31**(4):309-315

[51] Russell WR, Hoyles L, Flint HJ, Dumas ME. Colonic bacterial metabolites and human health. Current Opinion in Microbiology. 2013;**16**(3):246-254

[52] Vijay N, Morris M. Role of monocarboxylate transporters in drug delivery to the brain. Current Pharmaceutical Design. 2014;**20**(10):1487-1498

[53] Plöger S, Stumpff F, Penner GB, Schulzke JD, Gäbel G, Martens H, et al. Microbial butyrate and its role for barrier function in the gastrointestinal tract. Annals of the New York Academy of Sciences. 2012;**1258**(1):52-59

[54] Dörfel MJ, Huber O. Modulation of tight junction structure and function by kinases and phosphatases targeting Occludin. Journal of Biomedicine & Biotechnology. 2012;**2012**:1-14

[55] Ivanov AI, Parkos CA, Nusrat A. Cytoskeletal regulation of epithelial barrier function during inflammation. The American Journal of Pathology. 2010;**177**(2):512-524

[56] Ghanim H, Abuaysheh S, Sia CL, Korzeniewski K, Chaudhuri A, Fernandez-Real JM, et al. Increase in plasma endotoxin concentrations and the expression of toll-like receptors and suppressor of cytokine signaling-3 in mononuclear cells after a high-fat, high-carbohydrate meal: Implications for insulin resistance. Diabetes Care. 2009;**32**(12):2281-2287

[57] Ang Z, Ding JL. GPR41 and GPR43 in obesity and inflammation - protective or causative? Frontiers in Immunology. 2016;**7**:1-5

[58] Fukui H, Xu X, Miwa H. Role of gut microbiota-gut hormone axis in the pathophysiology of functional gastrointestinal disorders. J. Neurogastroenterol. Motil. 2018;**24**(3):367-386

[59] Oriach CS, Robertson RC, Stanton C, Cryan JF, Dinan TG. Food for thought: The role of nutrition in the microbiota-gut-brain axis. Clinical Nutrition Experimental. 2016;**6**:25-38

[60] Zhang M, Sun K, Wu Y, Yang Y, Tso P, Wu Z. Interactions between intestinal microbiota and host immune response in inflammatory bowel disease. Frontiers in Immunology. 2017;**8**:1-13

[61] Abreu MT. Toll-like receptor signalling in the intestinal epithelium: How bacterial recognition shapes intestinal function. Nature Reviews Immunology. 2010;**10**(2):131-143

[62] Hanke M, Keilian T. Toll-like receptors in health and disease in the brain: Mechanisms and therapeutic potential. Clinical Science. 2011;**121**(9):367-387

[63] McCusker RH, Kelley KW. Immune - neural connections: How the immune system's response to infectious agents influences behavior. The Journal of Experimental Biology. 2012;**216**(1):84-98

[64] Moyano-Porcile V, Olavarría-Ramírez L, González-Arancibia C,

Bravo JA, Julio-Pieper M. Short-term effects of poly(I:C) on gut permeability. Pharmacological Research. 2015;**101**: 130-136

[65] Kamada N, Seo SU, Chen GY, Núñez G. Role of the gut microbiota in immunity and inflammatory disease. Nature Reviews Immunology. 2013;**13**(5):321-335

[66] Milanski M, Degasperi G, Coope A, Morari J, Denis R, Cintra DE, et al. Saturated fatty acids produce an inflammatory response predominantly through the activation of TLR4 signaling in hypothalamus: Implications for the pathogenesis of obesity. The Journal of Neuroscience. 2009;**29**(2):359-370

[67] Haak BW, Wiersinga WJ. The role of the gut microbiota in sepsis. The Lancet Gastroenterology & Hepatology. 2017;**2**(2):135-143

[68] Thaiss CA, Zmora N, Levy M, Elinav E. The microbiome and innate immunity. Nature. 2016;**535**(7610):65-74

[69] Browning KN, Travagli RA. Central nervous system control of gastrointestinal motility and secretion and modulation of gastrointestinal functions. Comprehensive Physiology. 2014;**4**(4):1339-1368

[70] Foster JA, Rinaman L, Cryan JF. Stress & the gut-brain axis: Regulation by the microbiome. Neurobiology of Stress. 2017;**7**:124-136

[71] Manuscript A. Permeability in a region-specific manner. Neurogastroenterology & Motility. 2014;**25**(2):1-21

[72] Collins SM, Surette M, Bercik P. The interplay between the intestinal microbiota and the brain. Nature Reviews Microbiology. 2012;**10**(11):735-742

[73] Diaz Heijtz R, Wang S, Anuar F, Qian Y, Björkholm B, Samuelsson A, et al. Normal gut microbiota modulates brain development and behavior. Proceedings of the National Academy of Sciences of the United States of America. 2011;**108**(7):3047-3052

[74] Barbara G, Feinle-Bisset C, Ghoshal UC, Santos J, Vanner SJ, Vergnolle N, et al. The intestinal microenvironment and functional gastrointestinal disorders. Gastroenterology. 2016;**150**(6):1305-1318.e8

[75] Del Vecchio M, Procopio G, Cassata A, Pozzi P, Busto G, Celio L, et al. Stress-induced immune dysfunction implications for health. Tumori. 1999;**85**(1):6-11

[76] O'Farrell K, Harkin A. Stress-related regulation of the kynurenine pathway: Relevance to neuropsychiatric and degenerative disorders. Neuropharmacology. 2017;**112**:307-323

[77] Dantzer R. Role of the kynurenine metabolism pathway in Inflammation-induced depression: Preclinical approaches. In Inflammation-Associated Depression: Evidence, Mechanisms and Implications. Current Topics in Behavioral Neurosciences; Cham, Switzerland: Springer; 2016. pp. 117-138. ISBN: 978-3-319-51152-8

[78] De Punder K, Pruimboom L. Stress induces endotoxemia and low-grade inflammation by increasing barrier permeability. Frontiers in Immunology. 2015;**6**:1-12

[79] Schachter J, Martel J, Lin CS, Chang CJ, Wu TR, Lu CC, et al. Effects of obesity on depression: A role for inflammation and the gut microbiota. Brain, Behavior, and Immunity. 2018;**69**:1-8

[80] Smythe GA, Kapoor V, Kerr SJ, Guillemin GJ, Croitoru J, Brew BJ, et al. Kynurenine pathway metabolism

in human astrocytes: A paradox for neuronal protection. Journal of Neurochemistry. 2003;**78**(4):842-853

[81] Vécsei L, Szalárdy L, Fülöp F, Toldi J. Kynurenines in the CNS: Recent advances and new questions. Nature Reviews Drug Discovery. 2013;**12**(1):64-82

[82] Xu Z, Zeng W, Sun J, Chen W, Zhang R, Yang Z, et al. The quantification of blood-brain barrier disruption using dynamic contrast-enhanced magnetic resonance imaging in aging rhesus monkeys with spontaneous type 2 diabetes mellitus. NeuroImage. 2017;**158**:480-487

[83] Penninx BWJH. Depression and cardiovascular disease: Epidemiological evidence on their linking mechanisms. Neuroscience and Biobehavioral Reviews. 2017;**74**:277-286

[84] Unger JW, Livingston JN, Moss AM. Insulin receptors in the central nervous system: Localization, signalling mechanisms and functional aspects. Progress in Neurobiology. 1991;**36**(5):343-362

[85] Yahfoufi N, Alsadi N, Jambi M, Matar C. The Immunomodulatory and anti-inflammatory role of polyphenols. Nutrients. 2018;**10**(11):1618

[86] Cho J-H, Johnson GVW. Primed phosphorylation of tau at Thr231 by glycogen synthase kinase 3β (GSK3β) plays a critical role in regulating tau's ability to bind and stabilize microtubules. Journal of Neurochemistry. 2010;**88**(2):349-358

Section 2

Neuropathology

Chapter 5

Astrocytes and Inflammatory Processes in Alzheimer's Disease

Soraya L. Valles, Federico Burguet, Antonio Iradi, Martin Aldasoro, Jose M. Vila, Constanza Aldasoro and Adrián Jordá

Abstract

A significant increase in inflammation has been shown to be a crucial factor in the progression of the Alzheimer's disease (AD). Moreover, inflammatory signals are already present in mild cognitive impairment (MCI) patients before they develop AD. The amyloid hypothesis argues that in AD, there is an increase in oxidative stress caused by the accumulation of β-amyloid (Aβ) and that its elimination should be a priority. Also, hyperphosphorylation of the protein TAU occurs, which is characteristic of this disease. In AD oxidative stress processes occur and also inflammation. The basal chronic inflammation produces a cascade of cellular, such as astrocytes and microglial cells, and molecular processes in AD patients. We here have tried to explore the action of the inflammatory process and its implication in the neurodegenerative process of the AD. We can see that the role of Aβ is only one component that gives rise to inflammation, probably mediated by activation of microglia and astrocytes with the goal of getting rid of these brain waste products. In fact, it is related to a greater degree with the progression of the disease and worsening of the symptoms with the increase of phosphorylated TAU in different parts of the brain.

Keywords: astrocytes, microglia, neuroprotection, Alzheimer's disease, inflammation, oxidative stress

1. Introduction

Inflammation is a physiological process in response to various factors such as infection, trauma, and a long list of diseases that can promote it [1]. It is not uncommon to think that changes or failures that occur in their action mechanisms can lead to fatal consequences for humans. The inflammation originates because of a set of immune cells involved in the process that causes different changes in the inflamed area through signaling pathways composed of different groups of pro and anti-inflammatory molecules [2]. The resolution of the inflammatory process happens after the neutralization of the trigger. The cells of the immune system generate an anti-inflammatory activity, including lipoxins (for example, LXA4, RvE1) and cytokines such as interleukin-10 and interleukin-37, transforming growth factor-beta (IL-10, IL-37, TGF-β) [3]. Acute inflammatory processes will be resolved relatively quickly, while, however, resolution processes are not achieved in chronic inflammation [4].

IntechOpen

The differences between both types of inflammation, acute and chronic, reside at different levels. Regarding the cells involved in acute inflammation, neutrophils intervene in an infection context and eosinophils and mast cells in the case of allergies [5]. The chemical mediators involved in acute inflammation would be the complementary system, the kinins, the prostaglandins, the leukotrienes, the cytokines coming from several immune cells, and the gamma interferon of the T lymphocytes [6]. The lesions that are produced in this type of inflammation are itching, pus, and abscesses [7]. On the other hand, in chronic inflammation, we would have the participation of macrophages and lymphocytes mainly, which would produce cytokines as the main chemical mediators of this type of inflammation. As alterations, we would also have a rash (in the context of a cutaneous disease), and unlike the findings we had in the acute, in chronic we can have fibrosis and granuloma. These last two injuries are ultimately responsible for the effects of deterioration at central nervous system (CNS) and peripheral (SNP) level [8]. The study of neurodegenerative diseases excluded inflammation as an etiological agent of the disease. This was because there were no infiltrates of inflammatory cell similar to those that occur in infectious or autoimmune diseases [9]. Nowadays, there is an increasing amount of studies that position inflammation as being responsible for neurodegeneration through the participation of macrophages and the complementary system [10].

2. Specification of the process at the brain level

In the brain there is no reddening, local heat, or pain after acute inflammation. In the case of chronic inflammation in another organ, the participation of different immune cells takes place. But in the CNS, macrophages are essentially the representatives of the immune system [11].

In the CNS, the derivatives of tissue macrophages would be the microglia of the central nervous system. Microglia participate in numerous maintenance functions such as synapse management, neurogenesis, regulation of certain cognitive processes, and immunological protection [12]. Thus, the main hypothesis on the pathogenesis of Alzheimer's disease (AD) is that the plaques of β-amyloid (Aβ) and neurofibrillary tangles produce an acute inflammation in the brain, which activates these cells causing different inflammatory mediators, such as: proinflammatory cytokines, chemokines, macrophage inflammatory proteins, monocyte chemoattractant proteins, prostaglandins, leukotrienes, thromboxanes, coagulation factors, reactive oxygen species (and other radicals), nitric oxide (NO), complement factors, proteases, protease inhibitors, pentraxins, and C-reactive protein [13]. Due to the chemical composition of the Aβ plaques and neurofibrillary tangles, they stimulate a chronic inflammatory reaction with the intention of eliminating these brain structures [13]. Finally, this inflammatory reaction will produce a neuronal dystrophy mediated by the inflammatory mediators that are secreted by the microglial and astrocyte cells, as well as by the aggregates of amyloid fibrils [14].

3. Pathophysiology of Alzheimer's disease

The pathophysiology of Alzheimer's disease is very varied and there are different hypotheses on how it develops: the most accepted hypothesis in recent years was the amyloid hypothesis. The amyloid precursor protein (APP) will be able to be processed by either α-secretase, β-secretase, or γ-secretase. Depending on which enzyme does the app cut, we can have more or less neuroprotective profile; the α-secretase cleave produced a more neuroprotective one, while on the other hand, if the β- and γ-secretase participate sequentially, we will obtain metabolites that are harmful to neurons, producing greater amount of Aβ [13].

3.1 Role of α-secretase

α-Secretases are a family of proteolytic enzymes that adhere to APP in their transmembrane region. The secretases adhere to the fragment that, however, is processed by β-secretases and γ-secretases and that increases the β amyloid peptide [15]. These enzymes are members of the ADAM (disintegrin and metalloprotease domain) family that are expressed on cell surfaces. Furthermore, a metabolite by the action of secretase is APPsα, which has a not only neuroprotective action, but also neurotrophic effects have been observed and, therefore, neuroplasticity can be favored [16].

3.2 Role of β- and γ-secretases

The amyloid plaques are composed of a fragment of the APP: the 4-kD amyloid-β protein. The enzymatic processing of APP, resulting in Aβ, requires two enzymes: the γ-secretase, which is dependent on presenilin, and β-secretase, which is an aspartyl protease β-site APP-cleaving enzyme (BACE) (also known as Asp2, memapsin 2) [17, 18]. The BACE1 will function to split the APP, giving as result the βCTF (beta C-terminal fragment), which will later be cleaved again by γ-secretase to give rise to Aβ. On the other hand, this second excision could be caused by a mechanism different from that carried out by γ-secretase, which would be dependent on a 20S proteasome and whose malfunction would lead to an overproduction of Aβ in the same way [19].

3.3 The β-secretase: BACE

There are two BACEs, BACE1 and BACE2. BACE2 is a homolog discovered later than the enzyme BACE1 and shares 64% of similarity in its structure. By contrast, BACE2 is expressed at low levels in neurons and does not have the same activity against APP as BACE1 [20]. The BACE1 is doubly increased in the brains of patients with AD, compared to the brains of individuals without the disease. For this reason, it is considered that this enzyme is responsible for the initiation or acceleration of AD. Other studies show how BACE1 is also increased in response to stress: during oxidative stress, hypoxia ischemia, apoptosis, and brain trauma [18].

3.4 The γ-secretase

Research on the proteolytic processing of APP has provided information on the pathogenesis of Alzheimer's disease and on an unusual form of regulation of proteolytic processing within the domains of some membrane proteins, including APP, Notch, and ErbB4 [21]. Some of the enzymes responsible for α and β cleavage are already known. However, the molecular events that are involved in the cleavage produced by the γ-secretase, within the transmembrane domain of these proteins, are much more complex. Presenilins and nicastrin are necessary for this process. While the role of presenilins, in some cases, supports the idea that presenilins are found in the active site of the γ-secretase, other data indicate that they could have a more indirect function, as for example in the transport of substrates to the subcellular compartment for cleavage by the enzyme γ-secretase [22].

3.5 Role of β-secretase: BACE1 and γ-secretase in voltage regulation by sodium channel

The sodium channels Na1s are responsible for regulating for regulating the passage of Na^+ in the initial axonal fragments, Ranvier nodes, and neuromuscular junctions. These channels are formed by an α-subunit in the form of pore and two accessory β subunits which are transmembrane that modify the localization, surface cell expression, and inactivation of the alpha subunit by direct interaction, specially β2 subunit of the Na-1 channel that plays an important role since it undergoes degradation by BACE1 and γ-secretase [23]. These enzymes cleave an intracellular fragment of the C-terminal fraction that results in a transcription factor for Na1.1 mRNA and other protein levels, so that Na 1.1 levels accumulate intracellularly [23]. This fact explains the decreased expression of sodium channels on the surface of the hippocampal neurons of patients with AD, as well as in neuroblastoma cells producing BACE1, resulting in a lower sodium current density [23].

3.6 Differences between Aβ40 and Aβ42

To better understand the fact why Aβ42 promotes, to a greater extent, inflammation in AD than the Aβ40 peptide, it is necessary to emphasize its greater propensity to form amyloid plaques [24]. Studies performed by combining molecular dynamics and nuclear magnetic resonance (NMR) experiments with respect to the behavior of both peptides in water have shown that the Aβ42 peptide forms tangles more prominently [24, 25]. The differences that exist at the level of the chemical formula between the two peptides are only two amino acid residues at the C-terminus. However, at the level of biochemical and conformational interactions, there are clear differences [25]. In addition, while the N-terminal half presents a much smaller spectrum of possible conformations in its secondary structure, the C-terminal half of the Aβ42 peptide allows a greater number of possible conformations. Despite this, these studies showed that Aβ42 is more structured in water than Aβ40 [24]. Specifically, it is appreciated that the Aβ42 form has less flexibility than Aβ40 in its C-terminal half. This fact is produced by the formation of a beta hairpin in the sequence IIGLMVGGVVIA, involving short fragments of the structure between the residues of amino acids 31–34 and 38–41, reducing the flexibility in the Aβ42 peptide. Specifically, this must be the cause of the greater capacity to form amyloid plaques. On the other hand, a β-turn type VIB, centered on residues 35 and 36, is important for the alignment of the threads involved. In addition, the existence of hydrogen bonds between the pairs A30-A42, I32-V40, and L34-G38 adds stability to the structure of the beta fork [24, 25].

4. Identification and definition of the problem-question

In epidemiological studies of Alzheimer's disease, a significant increase in inflammation has been shown to be a crucial factor in the progression of the disease, as well as in the activation of microglia and in the increase of reactive astrocytes in these patients [26]. It should be noted that inflammatory signals are already present in mild cognitive impairment (MCI) patients before they develop AD [27]. In this study, we have tried to explore the action of the inflammatory process associated with Alzheimer's disease and its implication in the neurodegenerative process of the disease.

5. Interest of the review

Glial cells have a very important role in the protection of the central nervous system against damage and also in the repair of damaged nerve tissue [28]. Within the glia, astrocytes are the cell type prevalent in the brain [29]. Astrocytes increase neuronal viability and mitochondrial biogenesis, protecting neural cells from oxidative stress and inflammation induced by the toxic amyloid peptide [30–32]. Conversely, if chronic inflammation occurs, astrogliosis is triggered, produced by a reaction to inflammation and oxidative stress caused by toxic and inflammatory agents [33]. In Alzheimer's disease, complex changes and specific conflicts occur in different brain regions. The number of reactive astrocytes increases, engulfing and reducing the amyloid plaques. In addition, astrocytes surround the amyloid plaques and secrete proinflammatory factors, such as tumor necrosis factor (TNF) or interleukin 1 (IL-1) [34]. Currently, no hypothesis about what causes Alzheimer's disease has obtained favorable results. For years, it has been believed that the amyloid theory was the correct one and it was the most supported and financed by almost all the pharmaceutical companies around the world. The amyloid hypothesis argues that in AD, there is an increase in oxidative stress caused by the accumulation of Aβ and that its elimination should be a priority. There is a lot of research showing that increased levels of ROS have been linked to Alzheimer's disease [30, 35] but the effects of antioxidants in clinical studies have been disappointing either because high concentrations of antioxidants are pro-oxidants, or because the oxidative stress occurs relatively early in the course of the disease.

5.1 Mediators of the inflammatory process in AD

5.1.1 Cytokines

In AD, different cytokines have been detected, such as IL-1α, IL-1β, IL-6 and, similarly, higher amounts of the type B receptor IL-8 (IL-8RB) have also been found (in neurons in addition to the rest immune cells), unlike the type A receptor for IL-8RA that is only found in immune cells [36, 37]. It was already demonstrated that inflammatory signals are previously present in patients with mild cognitive impairment or with MCI before they develop AD [27].

The cytokine IL-1β constitutes one of the first secreted cytokines in response to lesions, as it is an important mediator of proliferation, differentiation, and apoptosis [38, 39]. The concentration of the said cytokine has been increased near the sites where the amyloid plaques are located [40]. More recently, it was observed that old mice had an increased basal neuroinflammation and they express IL-1β and IL-10 in the hippocampus compared to adult mice [41].

A study conducted in autopsies of 10 patients clinically diagnosed with AD showed that they had amyloid plaques and immunoreactivity for the cytokine IL-6. On the other hand, the control patients did not have immunoreactivity for IL-6 whether they presented plaques or not. From the plaques that were positive for the cytokine IL-6, it could be observed that they were most frequently found in diffuse plaques, less frequently in primitive plaques, and rarely found in compact and classic plaques [42].

5.1.2 Role of lipopolysaccharide (LPS)

The role of lipopolysaccharides in Alzheimer's disease has been studied by several research groups and it has been observed that treatment with these LPS induces chronic neuroinflammation [43, 44] and can contribute to deficits in learning and memory [44–46]. As previously known, LPS is an activator of microglia in the central nervous system and can induce a 2-fold increase in the expression of APP in the brains of mice with the Swedish mutation for APP [47]. In addition, it also caused an 18-fold increase in βCTF, suggesting an increased activity in turn of BACE1 and in turn an increase by up to three times in the amount of Aβ40 and Aβ42 [47]. While the previous study observed an increase in the brain in a non-specific manner, another study specifically analyzed the increase in glial fibrillary acidic protein (GFAP)-positive astrocytes in the cortex and hippocampus after treatment with LPS [48].

5.2 Alzheimer's disease as taupathy

5.2.1 Structure of the TAU protein

In electrophoresis gels, the TAU protein has been found in different isoforms depending on how the RNA has been processed and different levels of phosphorylation. This RNA is located on chromosome 17, it has at least 16 exons [49]. Other proteins besides tubulin have been described that can bind to the TAU protein: spectrin protein phosphatase 1, protein phosphatase 2a, presenilin 1, α-synuclein. Recent studies that have used mass spectrophotometry techniques indicate that it is more appropriate to measure the bacterially expressed MT-binding region (MTBR) domain of TAU, instead of the total TAU protein, this technic is more accurate to calculate the amount of TAU neurofibrillary tangles [50].

5.2.2 AD as taupathy

The TAU protein is a member of the microtubule-associated proteins (MAPs). The microtubules in the cells have a multitude of functions, among which we can highlight at the level of the neurons the formation of dendrites, axons, and their specific contacts [51]. Therefore, the TAU protein is necessary for the functioning and development of the nervous system and the presence of modified forms of the TAU protein gives rise to important pathological effects in the neurons that leads to neurodegeneration. Specifically in AD, phosphorylation of TAU protein is produced by glycogen synthase 3ß (GSK3) [49]. This TAU protein is abnormally phosphorylated and will form the neurofibrillary tangles in the neuronal cytoplasm, constituting one of the most important histological features of Alzheimer's disease. As previous works demonstrated, the number of these balls will be directly related to the severity of the symptoms of the disease [52]. The structure of these microtubules will be formed by double helix subunits that are intertwined with levorotatory filaments that are composed of the following proteins: intermediate filaments, neurofilaments of medium and high molecular weight; proteins associated with microtubules MAP2 and TAU; actin; and ubiquitins [53, 54], which show characteristics different from normal neurofilaments and normal microtubules.

In 1995, a study in autopsies done with eight patients with diagnostic criteria for Alzheimer's disease and six control patients of similar ages indicated important changes between TAU and inflammation. The brain of these 15 subjects was extracted without exceeding 15 h of postmortem and, later, samples were taken from the hippocampus; from the frontal, temporal, and occipital lobes; and from the cerebellum. AD patients presented a direct relationship between higher concentrations of the activated IL-α and higher load of neuritic plaque TAU2+ (TAU 2-immunoreactive). There is a strong association between the presence of IL-1α +, microglia, and TAU protein plates in patients with Alzheimer's disease [55]. Recently, it was observed in a microglial culture model together with neurons that the inflammatory response mediated by LPS-induced microglia leads to

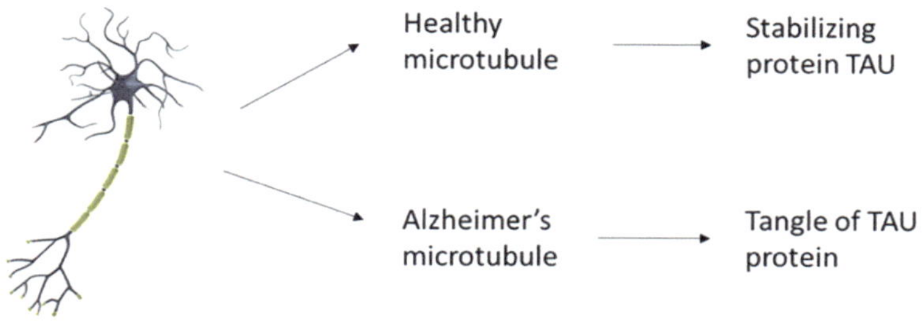

Figure 1.
TAU protein in health and Alzheimer's disease.

hyperphosphorylation of TAU mediating the greater kinase of the TAU protein, GSK3β (kinase glycogen synthase kinase 3β) [56]. On the other hand, chronic inflammation causes phosphorylation of TAU and worsens pathology in neurons that express many inflammatory receptors and molecules, including, MHC-I, TNFR1, IL-1R, and TLR. As a result of this, it allows them to interact directly with microglia [57]. Inflammatory signals can consequently directly activate neuronal protein kinases and phosphatases, such as cyclin-dependent kinase 5 (CDK5), glycogen synthase kinase-3β (GSK3β), ERK, and protein-phosphatase 2A (PP2A), which regulates phosphorylation of TAU and the assembly of neuronal microtubules [43, 44, 58, 59] (**Figure 1**).

TAU, and its function regarding neuronal and microglial interactions in brain immune chain reactions, as in AD and its progression, could be initiated, by age-related chronic inflammation. There is an increase in scientific evidence suggesting the importance of the mechanism in the synaptic pruning regulation, neurogenesis, immunological chain reaction-mediated cognitive functions in brain cells, and LTP, even if its complete relevance is still to be confirmed [44].

The mechanisms of inflammation effects on TAU and its pathological influx remain constant even if broad investigations have been carried out on Aβ and inflammation pathway. Persistent microglial activity and inflammation are established to be the causes of a broad release of TAU sub-species [44–47]. As for the mechanisms of TAU-induced inflammatory responses leading to pathology, several sources point to an acceleration of the onset of main protein kinases, which take care of the phosphorylation of the TAU protein. Microglia-perpetuated liberation of TNF-α has proven to provoke the accumulation and aggregation of TAU in in vitro neurons [48]. On the other hand, blocking microglia with minocycline reduces the inflammatory response and propagation of pathology related to TAU in experiments performed with hTau mouse models [60]. Moreover, the inhibition of inflammation by arginase-1 overexpression counteracts the activity of nitric oxide synthases, and facilitates autophagy and the decrease of TAU pathology in the TAU-transgenic mouse model rTg4510 [61]. A process that has been reported as important in stress-induced mechanisms and which can be genetically suppressed by the corticotropin-releasing factor receptor is the stimulation of toll-like receptor 4 (TLR4), which increases GSK-3β and CDK-5, which phosphorylate TAU [62].

To perform the "synaptic pruning" by the microglia, a cytosine secreted by the neural cells called fraktalkine (CX3CL1) that is excreted in large quantities in the brain compared to the rest of the organs of the body is needed [63]. Its receptor CX3CR1 is expressed in large quantities by microglia [64]. Previously, it was demonstrated that neuroinflammation via the receptor deficiency for fractalkine (CX3CR1) promotes taupathy and neurodegeneration in mouse models in which systemic inflammation mediated by LPS had occurred. First, Mapt$^{+/+}$ neurons showed high levels of Annexin V (A5) and TUNEL (markers of neurodegeneration) when they were grown together with microglia Cx3cr1$^{-/-}$ treated with LPS. Second, a population of positive neurons for TAU protein phospho-S199 (AT8) in the dentate gyrus is also positive for (CC3) for mice treated with Cx3cr1$^{-/-}$. Third, the genetic deficiency of TAU in Cx3cr1$^{-/-}$ mice resulted in reduced microglial activation, which altered the expression of inflammatory genes in those neurons positive for CC3 compared to Cx3cr1$^{-/-}$ mice [44]. These results suggest that pathological changes in TAU mediate the neurotoxicity induced by inflammation, while Mapt deficiency is neuroprotective. It was proposed that this earlier phenomenon was probably associated with the indirect reduction of microglial activity due to the decrease in the production of pathological species of TAU, observed in a transgenic mouse model rTg4510, which expresses the mutation in P310L (4R0N TauP301L) and initiates taupathy within 3–5 months. Brain stimulation of TLR4 by LPS in the aforementioned mouse model also produces activation of microglia and phosphorylation of TAU [65]. In another investigation using the 3xTg-AD transgenic mouse model, which develops both Aβ and taupathies, chronic treatment with LPS results in phosphorylation of CDK5-dependent TAU without affecting Aβ levels in adult animals (~6 months old). TAU phosphorylation was observed by immunohistochemistry techniques when treated with LPS and PBS samples of the aforementioned 3xTg-AD mice by two tests: in the first one in the Ser202/Thr205 residues that were recognized by AT8, they presented up to twice as much AT8 activity in the samples treated with LPS as those that were administered PBS; the second test detected that in the Thr231/Ser235 region, recognized by AT180, there was more activity this time of AT180 in the presence of LPS. However, the same did not happen in the Ser396/Ser404 region that was recognized by PHD finger protein 1 (PHF-1), where the sample with LPS was not altered to a greater extent compared to that which was administered PBS [66]. TLR4's activation has proven to initiate the TAU-mediated pathologies in a more powerful manner in aged 3xTg-AD mice (more than 12 months of age), which means that the influence of TAU over inflammatory mechanisms grows stronger with age. Older groups of 3xTg-AD, which received a chronic LPS treatment, showed TAU phosphorylation in AT8, AT180, and PHF-1 epitopes, as well as TAU accumulation and aggregation as neurofibrillary tangles and cognitive deterioration, appearing, though, no changes in platelet saturation of Aβ. In this tested, aged animals, TAU pathology modulation induced by TLR4 is principally dominated by GSK3β (glycogen synthase kinase-3ß), the latter data were verified through the inhibition with lithium of GSK3β, where a reduction was observed of the phosphorylation of TAU and the accumulation in its insoluble form together with the reversal of memory problems [67].

Another possible route deduced in a study done in the brains of patients with early onset of Alzheimer's disease (FAD) with the Swedish mutation for APP, corticobasal degeneration (CBD), and progressive supranuclear palsy (PSP) three well-known taupathies, which presented PS3 positive vesicles in the frontal cortex, which indicates that autophagic vesicles accumulated in the said location. In addition, LAMP1 (lysosomal-associated membrane protein 1) lysosomal markers were found in FAD and CBD, and cathepsin in the three mentioned diseases. Thus, this study presents a possible role of the autophagy-lysosome pathway that would contribute to the development of primary taupathies as well as FAD [68]. The unbalanced increase in IL-1β expression in 3xTg-AD models generated inverse effects in amyloid-based pathologies and TAU accumulation, by increasing the addition of its pathological forms while decreasing the total quantities of Aβ plaques. The elimination of such plaques is powered up by the effects of IL-1β in an increase in Aβ plaque surrounding activated microglia. This process also augments the proinflammatory status, in a directly proportional intensity to age, by means of its elimination. In turn, it was also found in this experiment that it generates the activation of GSK3β and p38-MAPK, which leads to a higher level of phosphorylation in TAU [69].

It has been demonstrated in an experiment carried out in 3xTg-AD transgenic mice that the inhibition of IL-1 signaling decreases the activation of the kinases CDK5/p25, GSK3β, and p38-MAPK, as well as reduces the phosphorylation levels of TAU [66]. On the other hand, the blockade of IL-1R showed that it altered the inflammatory responses of the brain (related to a lower activity of NF-κB), reduces cognitive deficits, and notably attenuates the pathology attributed to TAU, and decreases the oligomeric and fibrillary forms of Aβ. Similarly, it was found that there was a reduction of the cytokine derived from astrocytes, S100β, and in neuronal signaling with Wnt/β-catenin in 3xTg-AD brains [66, 70]. In addition to the complex connection between inflammation and AD, it has been shown that opposite effects can be seen in Aβ and TAU produced with inflammation. For example, it was shown that the main risk factor for Alzheimer's disease, aging, seems to cause a decrease in the levels of sirtuin 1 (SIRT1), which is related to microglial aging. Thus, this deficiency in the microglial SIRT1 with age results in an excessive production of IL-1β, which in turn causes pathology through TAU in addition to cognitive deficits. The deficiency of microglial SIRT1 induces a hypomethylation of specific loci CpG in the promoter for IL-1β, with elevation of IL-1β transcription [71].

Parallel studies affirm what was previously stated. For example, CX3CR1 deficiency in mouse models of amyloidosis mitigates the accumulation of Aβ by altering microglial activation and promoting microglial phagocytosis [65, 72]. On the other hand, blockade of CX3CR1 signaling increases IL-1β/p38-MAPK-mediated TAU phosphorylation in the hTau taupathy model [43]. The genetic suppression of CX3CL1 anchored to the membrane, ligand of CX3CR1, in models of amyloid pathology and taupathy in the APP/PS1 mouse models also reduces the deposition of Aβ through the increase of phagocytosis mediated by microglia and at the same time induces phosphorylation of neuronal TAU [73], thus having similar effects as in the deficiency of the microglial receptor CX3CR1, as shown above. In addition, it was already studied that a loss of function was mediated by mutations of progranulin, which has been associated with frontotemporal dementia [74], and results in an increase in the activation signal of tyrosine kinase binding protein TYRO (TYROBP) and Aβ microglial phagocytosis in the APP/PS1 mouse model, while TAU pathology increases in mice expressing the human TA30 PIL mutation [75]. Obviously, these opposite effects induced by the immune signal in the accumulation of Aβ and TAU raise concerns as to the direction of the therapies relating to mitigate one or both of these effects by activating or inhibiting inflammation in the context of Alzheimer's disease. As we have seen in previous experiments, it is already known clinically and is explicitly stated in a research that TAU levels correlate better with cognitive deficits observed during the disease process [76]. The development of strategies to modulate the immune system to act in the deposition of Aβ and TAU hyperphosphorylation will probably produce better clinical results.

5.2.3 Astrocytes and inflammation

Analogous to microglia, astrocytes play multiple roles in the organization and maintenance of brain structure and function. Multiple studies show that astrocytes dynamically modulate information processing, signal transmission, neural and synaptic plasticity. As well as, homeostasis of the blood-brain barrier, and its role in immune responses. The evidence shows us how during cerebral ischemia, it acts as a protector, whereas against inflammation mediated by the lipopolysaccharide of *Escherichia coli*, its intervention seems to be harmful [77]. In the cells of the retina, however, it has been proven that through the production of lipoxins, it has

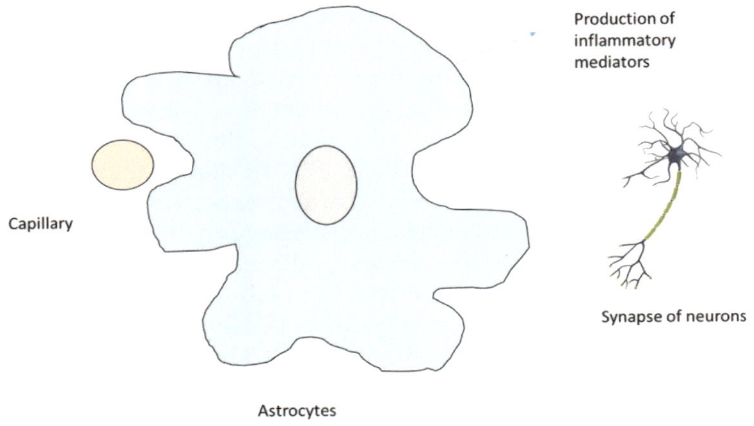

Figure 2.
Implication of astrocytes in inflammation.

an anti-inflammatory and neuroprotective effect against acute and chronic lesions [78]. Similarly, the role of the cytokine IL-33 produced by astrocytes has recently been demonstrated for the microglial approach to the synaptic terminals, as well as the development of neural circuits [79]. In previously mentioned studies describing the action of IL-1α^+, it is concluded that there is also a correlation between IL-1α and the greater number of GFAP$^+$ astrocytes (GFAP-immunoreactive astrocytes) [80]. On the other hand, it has been demonstrated in an experiment carried out in mice with multiple sclerosis TNF-α alters synaptic transmission and produces interferences at the cognitive level [81]. Other studies have shown that the activation of certain transcription factors are also involved, developing protective effects (STAT3) [82] or injurious effects (NF-κB) [83] (**Figure 2**).

5.2.4 Role of astrocytes in amyloid production

The role of astrocytes in the amyloidogenic pathway is currently being widely studied. For a long time, it was thought that neurons were the only type of cell that expressed high levels of BACE1 and, therefore, that neuron was the only type of cell capable of producing Aβ [84]. However, studies have shown that astrocytes express BACE1 at sufficient levels to generate Aβ, and that expression can be increased by cell stress [85–89]. In addition, stressors can upregulate the expression of APP and, therefore, the secretion of Aβ. In contrast, the effect of cellular stress on the activity of γ-secretase in astrocytes has not yet been fully clarified.

The production of Aβ will lead to activation of the microglia and astrocytes in order to get rid of these brain waste products [90–92]. Similarly, genetic studies have identified polymorphisms of a single nucleotide in inflammatory genes that are associated with the risk of AD, highlighting the role of inflammation in AD [86, 93–95]. In addition, it has been observed that patients with Alzheimer's disease have more proinflammatory cytokines and activated inflammasomes [96]. As demonstrated in studies that claim an increase in both glial fibrillary acid protein (GFAP) and S100β expression, they lead to greater astrogliosis in postmortem tissues of human patients and experimental models in mice. In the same way, a correlation has been found, in different studies, between the degree of astrogliosis and cognitive deterioration [32, 96, 97]. As astrocytes substantially exceed the number of neurons in the brain, the identification of cellular environment factors (such as inflammation), which promote the production of astrocytic Aβ, could redefine our therapeutic targets when it comes to fighting Alzheimer's disease.

5.2.5 S100β and inflammation

The cytokine S100β is known to be an important neurotrophic agent during fetal development, both in neuroblasts and in the glia [98]. In addition to this known function, it is known that it directly contributes to the activation and subsequent gliosis, stimulating the proliferation of astrocytes and inducing morphological changes [70]. Furthermore, the IL-1 produced in the microglia, is the responsible for the overproduction of S100β.

The distribution of S100β contained in activated astrocytes by ELISA and immunohistochemistry was studied, as shown by many, few, or no neuritic plaques in the context of Alzheimer's disease. Postmortem samples were obtained from both patients diagnosed with AD and control patients from the hippocampus, temporal lobes, frontal lobes, occipital lobes, brain stem, and cerebellum. The results indicated that the density of cells that were S100β$^+$, identified with activated astrocytes, was higher around the neuritic plaques in certain areas of the brain. By order, the concentration was found to be more remarkable in the hippocampus > temporal lobe > frontal lobe > occipital lobe > protuberance, and no neuritic plaques were found in the cerebellum. The importance of these results lies in the fact that the regulatory role of the cytokine S100β contributes to the development or maintenance of dystrophic neurites observed in neuritic plaques. Furthermore, overexpression of S100β shows that it has been related to a higher degree of dysfunction and neural loss in AD caused by an intracellular increase in calcium levels [70].

5.2.6 Astrogliosis

Astrogliosis occurs in the presence of a central nervous system lesion. Inflammatory mediators made by microglia, neurons, oligodendrocytes, endothelial cells, leukocytes, and other astrocytes initially cause astrocytes to become reactive [77]. To better understand the process of astrogliosis, we must bear in mind that a series of changes occur at the phenotype level of astrocytes, which induce a specific expression. This was demonstrated in an experiment using arrays (Affymetrix GeneChip arrays) to define the genetic expression of different populations of reactive astrocytes isolated at different time periods using two models of injury (neuroinflammation and ischemic stroke) in mice. It was observed that this reactive gliosis had a rapid, but rapidly diminished, pattern of induction of gene expression after damage, where Lcn2 and Sertapina3n were identified as the major markers of reactive astrocytes. It was also seen that the pattern of expression experienced during ischemic stroke had a protective profile, whereas in the population of mice in which neuroinflammation was induced by the use of LPS, it turned out to be, on the contrary, detrimental [77]. Moreover, using high-density microarray, reactive astrocytes also produced detrimental effects (in vitro models from multiple sclerosis, neoplasms and stroke), and was identified up to 44 different transcription patterns present in the different pathological models mentioned [99].

In astrocytes, the first morphological change is the process of hypertrophy that is intimately related to the greater expression of intermediate filaments, attributed to the action of GFAP [99]. Although the consequences of GFAP expression are not fully understood, it is known that they have a determining role in limiting the creation of Aβ plaques. The impact of this reactive astrogliosis is complex: reactive astrogliosis can be both harmful or beneficial at the time the cells are affected. Reagent astrocytes will surround the Aβ plaques and will express receptors such as receptor for advanced glycation end products (RAGE), receptor-like LDL protein (low-density lipoprotein), membrane-associated proteinglycans, as well as receptor-like scavenger receptors to bind to Aβ [100]. Reactive astrocytes will be neurotoxic when they generate reactive oxygen species or proinflammatory cytokines [101]. In order to understand the role of cerebral gliosis, the balance between the mechanisms that orient toward the neuroprotective or neurotoxic effect must be taken into account.

Patients with AD showed reactive astrocytes as shown by PET images [102, 103] and also, before the formation of plaques in transgenic APP mice [104]. Reactive astrocytes, depending on the level of gliotransmitters (including glutamate, ATP, serine-d and GABA) can produce inhibition of neuronal activity [105]. There is a consensus that the role of GABA is to protect neuronal cells in the brain [106]. In the amyloid plaques, an increase in the GABA protein has been detected in the reactive astrocytes that surround the plaques and that cause a greater release in the extracellular space [105]. It has been studied that these investigations have their limitations, since normally studies are carried out in mouse models, while in the human species there are many more processes to take into account [107].

5.2.7 Astrocytes, chemokines, and cytokines

Astrocytes can sometimes release reactive oxygen species (ROS), chemokines, or cytokines (CCL3, CCL4, CCL1, IL-1, for example) [108, 109]. Normally, those responsible for expressing these substances are going to be the so-called reactive astrocytes that cause functional changes by the expression of genes and the formation of glial scars that can be beneficial [81] or harmful to cells [82]. By using lipopolysaccharide (LPS) as an inducer, astrocytes increase the expression of many genes (C3a, C3b, C5, lectin) in the complement cascade that can be harmful [82]. On the other hand, it has been shown that positive regulation of trophic factors after ischemic damage is a protective mechanism [81]. Following the same line, inflammation is an essential factor in the progression of Alzheimer's disease in humans, demonstrating that this inflammation promotes the activation of microglia and an increase in reactive astrocytes that change their shape and increase the ramifications to go to the place of injury [110].

Relating astrogliosis to inflammation, both resting astrocytes and reactive astrocytes can secrete numerous cytokines capable of inducing inflammation, such as IFNγ, IL-1β, TNFα, IL-6, and TGFβ [37, 111–113]. IFNγ is a potent regulatory cytokine that activates microglia and promotes inflammation in the brain and is overproduced in the brains of patients with AD [114] both by microglia and astrocytes, despite which it is produced in the first instance by T cells [115, 116]. On the other hand, TNFα is a cytokine involved in the acute phase of inflammation and is also elevated in the serum, cerebral cortex, and cerebrospinal fluid of patients with AD [117]. In a study conducted by scientists at the Rostkamp Institute of the Department of Psychiatry at the University of South Florida, it was demonstrated in mice that those which were deficient in CD40, which is a gene that codes for the receptor TNF (Tumor Necrosis Factor), had a reduced activity of BACE, Aβ, and gliosis in comparison to the samples that presented normal quantities of CD40 [118]. IL-6 can have both proinflammatory and anti-inflammatory effects and has also been found elevated in plasma, cerebrospinal fluid, and in the brains of Alzheimer's patients [39, 119–122]. IL-1β constitutes one of the first cytokines secreted in response to lesions, as it is an important mediator of proliferation, differentiation, and apoptosis. The concentration of the said cytokine has been increased near the sites where the amyloid plaques are located [38–40].

A specific polymorphism in the transforming growth factor β1 (TGFβ1), an immunosuppressive cytokine, is also related to the risk of developing AD [123]. In addition, postmortem brains analyzed from Alzheimer's patients contained higher levels of TGFβ, specifically in their plaques, suggesting their involvement in the same disease [124, 125]. Other studies performed in older mice that overexpress TGFβ in astrocytes promoted the deposition of Aβ, and those astrocytes containing TGFβ1 were located in the vicinity of the Aβ deposits in those mice that overexpressed APP with Swedish mutation [126–129]. Finally, astrocytes release purines that can influence the development of AD and activate the production of inflammatory proteins, decreasing anti-inflammatory proteins [108].

5.2.8 Astrocytes and expression of APP

As it has been already mentioned before, APP is the substrate prior to Aβ after erroneous processing by the BACE1 and γ-secretase enzymes. The expression of APP by astrocytes has been demonstrated by the identification of APP695, APP751, and APP770 mRNAs found in non-neuronal cells [126] and in rat astrocytes [130]. In addition, it has been shown that multiple proinflammatory cytokines upregulate APP in both mouse and human brains (investigating neuroblastoma cells and other non-neuronal cells such as human astrocytes) [131]. These findings imply that neuroinflammation in reactive astrocytes expresses higher levels of APP than when mice are at rest and they, therefore, may end up producing more β amyloid. Similarly, in APP/PS1 mice, an increase in chemokines and their receptors, compared to wild type mice, such as CCL3, CCL4, CCL1 and the receptors CCR5 and CCR8 was detected [108].

Several studies have shown that the transcription factor for APP (AP-1) is found in the promoter region of many of the acute phase proteins of inflammation that are induced by the cytokines IL-1β and IL-6, suggesting that the expression of APP is regulated in the same way by these specific cytokines [132, 133]. Moreover, astrocytes stimulated with different combinations of cytokines (LPS + IFNγ, TNFα + IFNγ, and TNFα + IL-1β + IFNγ) increased the expression of APP [89].

5.2.9 Astrocytes and cancer

The type of tumor and its location are determined by age; for example, infratentorial astrocytoma and midline tumors, such as medulloblastoma and pinealoma, anaplastic astrocytoma, and glioblastoma predominate in adulthood [134]. Although meningiomas are the most frequently detected in the series of autopsies, glioblastomas are the most frequently detected in the brain. Some brain tumors such as schwannoma, sarcoma, glioma, and meningioma are detected after the patient has been exposed to cancer therapy with chemotherapy and/or radiotherapy. Until now it was thought that only glial cells and stem cells were responsible for the emergence of glioblastoma, but it is now known that mature neurons can also induce this type of cancer. This is due to the fact that these cells revert to an undifferentiated state that is directed to proliferate as an uncontrolled tumor [135].

Radial glia are stem cells that develop from a progenitor stem cell in the embryo and adult brain [136]. The neuroblastoma cells are radial glia or precursors of astrocytes that can develop before their differentiation into neurons. In the same way, glial cells can also develop different types of cells besides neurons such as oligodendroglia and astrocytes [137]. All these types of cells can turn into cancer and affect the normal function of the brain. Then, astrocytes and their progenitor cells can cause cancer and destroy many functions in the brain. It is interesting to note that in some astrocytomas, the patients increase their cognitive capacity, memory, and spatial vision, before the disease begins and also when the cancer is present [138], which makes us think and throws more evidence to the role of astrocytes in modulating cognitive brain functions or memory.

5.3 Protective role of astrocytes

5.3.1 Oxidative stress, AD, and the protective role of astrocytes against oxidative stress

Hydrogen peroxide (H_2O_2), superoxide (O_2^-) and hydroxyl radicals (OH^-) are the aforementioned reactive oxygen species (ROS). Due to the rate of oxidative metabolism, the SNC is especially susceptible to the damage suffered by them [139]. Under stable physiological conditions, the homeostasis of ROS is under control and this is crucial for the proper organic functions. ROS stimulates the proliferation of brain cells, but at high concentrations, ROS has harmful effects on different cellular structures such as membranes, DNA, and enzymes, which can lead to cell death [140]. The reduction of molecular oxygen is not complete in the respiratory chain, producing ROS continuously and thus affecting different cellular components such as proteins or lipids [141]. To return to the state of physiological equilibrium, the brain has several enzymes, such as peroxidase, superoxide dismutase (SOD), oxidase, and NADPH oxidase (NOX). Neurons have fewer defenses against ROS than astrocytes and cooperation between them is important for neuronal resistance against ROS [30, 142, 143]. Astrocytes contribute to the survival of neurons by detoxifying the ROS enzymes (GSH peroxidase and catalase), increasing antioxidant proteins (GSH or glutathione, vitamin E and ascorbate) and the biogenesis of mitochondria and reducing the activity of metals which can produce redox [31, 144–146]. The most powerful antioxidant protein in the brain is GSH produced by astrocytes and neurons, but neurons depend on astrocytes because they do not use extracellular cysteine efficiently and, therefore, need astrocytes to supply it. In addition, with respect to ascorbic acid, another important antioxidant in the nervous system, we depend on diet to obtain it [147].

Ascorbic acid is released by the astrocytes in the extracellular space and is absorbed by the neurons, where thanks to ascorbate the formation of ROS diminishes and its oxidized form is converted to be recovered by the astrocytes and converted again to ascorbic acid [148]. In addition, the lactate shuttle between astrocytes and neurons is favored by ascorbic acid [149]. Changes in ascorbic acid homeostasis are actually involved in different neurodegenerative diseases and have been analyzed for the treatment of diseases, such as Parkinson's and Huntington's disease [148]. In addition, astrocyte prevention in redox production caused by active metals has been demonstrated as a result of the ability to sequester metals by this cellular type [144].

The increase in ROS levels is related to AD [35], but the effects of antioxidants in clinical studies have been disappointing because the high concentration of antioxidants acts, in many cases, as pro-oxidants. It may also be due to the fact that oxidative stress occurs relatively early in the course of AD and therefore, by its administration at later stages, no results are obtained, or else that the combination of antioxidants does not work in clinical situations in humans [150]. As already shown, astrocytes protect neurons from oxidative stress, producing antioxidant proteins. The toxic amyloid beta peptide causes the production of hydrogen peroxide by astrocytes [151], as shown previously [30], and they release ROS in response to beta amyloid through the pentose-phosphate pathway [151]. In addition, in patients with Alzheimer's disease, there is a fall in the brain cleansing process produced by astrocytes during the sleep period. On the other hand, Haydon showed that the sleep/wake cycle is modulated by astrocytes and is also altered in AD [152]. This finding also demonstrates the close relationship between astrocytes and Alzheimer's disease.

After demonstrating the important role of astrocytes in protecting neurons from oxidative stress, we can deduce that all those conditions mentioned where the astrocyte undergoes changes in its function, beyond the strictly physiological, will result in poor protection of the said neurons and the rest of brain structures in front of different harmful agents and neuronal damage.

5.3.2 Prejudicial and protective role of astrocytes

As we have seen previously, the role of astrocytes is essentially protective. This was also demonstrated by another group finding a mechanism different from those previously studied by cytokines and other inflammatory agents, in which it was shown that the astrocytes surrounding the plates increase the release of ATP in transgenic APP/PS1 mice and this happens because the Ca^{2+} concentration increases within the cell [153]. This last fact gives us the idea that an increase in ATP in astrocytes and neurons could help to reduce the neuronal death that occurs in Alzheimer's disease. The increase in the production of ATP by the mitochondria of astrocytes could help to recover and reduce the development of the disease [153]. The neurotransmitter glutamate is released by astrocytes in the presence of Aβ and can cause neuronal loss as well as synaptic damage by activation of NMDA receptors [154, 155]. In addition, astrocytes release purines that can influence the development of AD and activate the production of inflammatory proteins, decreasing anti-inflammatory proteins, such as PPAR-γ [108, 156]. This is probably due to the effect of reactive astrogliosis that may have beneficial effects [82] or detrimental effects [83] for neurons, and because these two different reactions depend on the type of triggering of the astrogliosis. Nevertheless, in our laboratory, we demonstrated that astrocytes play a significant role in neuron protection. Astrocytes promote neural viability and improve oxidative stress defense mechanisms with anti-inflammatory effects against $A\beta_{1-42}$ peptide toxicity. It is probable that the protective effects of astrocytes are related with the mitochondrial biogenesis (**Figure 3**). This could be a complex epigenetic process in Alzheimer's disease pathogenesis [30].

In conclusion, we can see that the role of Aβ, which had been an essential pillar in the etiopathogenesis of Alzheimer's disease for decades, is only one component that gives rise to inflammation, probably mediated by activation of microglia and astrocytes with the goal of getting rid of these brain waste products, although this effect has already been shown to be produced in the same way by different

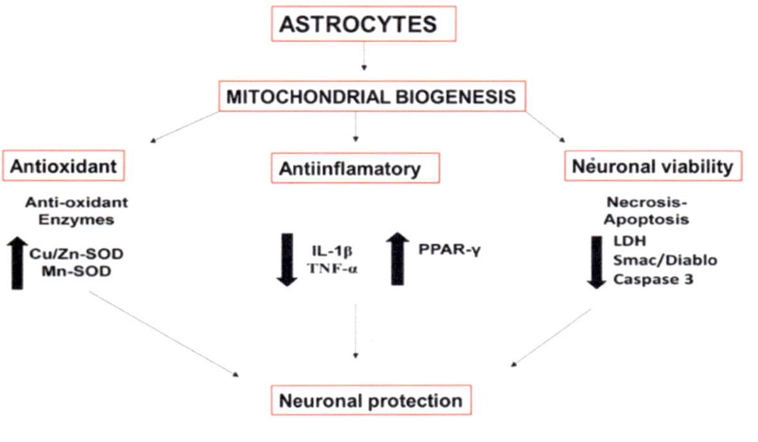

Figure 3.
Protective effects of astrocytes.

mediators. In fact, it is related to a greater degree with the progression of the disease and worsening of the symptoms with the increase of phosphorylated TAU in different parts of the brain. In the last years, the therapies have been focused on elimination of the Aβ from the brain of the Alzheimer's patient with poor results [157]. In addition, reactive astrocytes greatly increase NRF-2, which is an antioxidant protein and could produce beneficial effects in Alzheimer's disease [157]. The regulation of oxidative stress or inflammation could help the conservation of neurons located near astrocytes and microglia. Future therapies should be aimed at the development of specific drugs that control the formation of reactive astrocytes and that favor the correct resolution of the inflammation produced by Alzheimer's disease. The study of the genetic mechanisms that predispose to increase amounts of hyperphosphorylated TAU or those that decrease phosphorylation of TAU would be interesting in order to understand cellular mechanisms implicated in AD [157]. Furthermore, the study of the main trigger of this basal chronic inflammation that worsens the clinical symptoms of AD patients, should be crucial to find new therapeutic strategies. Finally, regarding the relationship that exists between the astrocytes and the cells of the nervous system, there would be a greater study of the functions of these cells in the healthy individual. The control of the mechanisms and the understanding of the relationship between astrocytes with other neural cells could help, in the same way, to the therapy of Alzheimer's disease.

Abbreviations

AD	Alzheimer's disease
IL-1β	interleukin 1β
TNF-α	tumor necrosis factor α
BACE	aspartyl protease β-site APP-cleaving enzyme
Aβ	β-amyloid
ADAM	disintegrin and metalloprotease domain
LXA4	lipoxin A4
IL-10	interleukin 10
IL-37	interleukin 37
TGF-β	transforming growth factor-beta
CNS	central nervous system
PNS	peripheral nervous system
APP	amyloid precursor protein
βCTF	beta C-terminal fragment
NO	nitric oxide
MCI	mild cognition impairment
ROS	reactive oxygen species
LPS	lipopolysaccharide
MAP	microtubule-associated proteins
CX3CL1	fraktalkina
GSK3β	glycogen synthase kinase-3 β
NF-κB	nuclear factor κB
SIT1	sirtuin 1
TLR4	toll like receptor 4

Author details

Soraya L. Valles*, Federico Burguet, Antonio Iradi, Martin Aldasoro, Jose M. Vila, Constanza Aldasoro and Adrián Jordá
Department of Physiology, School of Medicine, University of Valencia, Spain

*Address all correspondence to: lilian.valles@uv.es

References

[1] Ren Z, He H, Zuo Z, Xu Z, Wei Z, Deng J. The role of different SIRT1-mediated signaling pathways in toxic injury. Cellular & Molecular Biology Letters. 2019;**24**:36

[2] Cho K. Emerging roles of complement protein C1q in neurodegeneration. Aging and Disease. 2019;**10**(3):652-663

[3] Fullerton JN, Gilroy DW. Resolution of inflammation: A new therapeutic frontier. Nature Reviews. Drug Discovery. 2016;**15**(8):551-567

[4] Bollen J, Trick L, Llewellyn D, Dickens C. The effects of acute inflammation on cognitive functioning and emotional processing in humans: A systematic review of experimental studies. Journal of Psychosomatic Research. 2017;**94**:47-55

[5] Diwakar L, Cummins C, Lilford R, Roberts T. Systematic review of pathways for the delivery of allergy services. BMJ Open. 2017;**7**(2):e012647

[6] Serhan CN. Treating inflammation and infection in the 21st century: New hints from decoding resolution mediators and mechanisms. The FASEB Journal. 2017;**31**(4):1273-1288

[7] Kumari S, Pasparakis M. Epithelial cell death and inflammation in skin. Current Topics in Microbiology and Immunology. 2017;**403**:77-93

[8] Khiroya R, Macaluso C, Montero MA, Wells AU, Chua F, Kokosi M, et al. Pleuroparenchymal fibroelastosis: A review of histopathologic features and the relationship between histologic parameters and survival. The American Journal of Surgical Pathology. 2017;**41**(12):1683-1689

[9] Rose NR. Prediction and prevention of autoimmune disease in the 21st century: A review and preview. American Journal of Epidemiology. 2016;**183**(5):403-406

[10] Chitnis T, Weiner HL. CNS inflammation and neurodegeneration. The Journal of Clinical Investigation. 2017;**127**(10):3577-3587

[11] Franco R, Fernández-Suárez D. Alternatively activated microglia and macrophages in the central nervous system. Progress in Neurobiology. 2015;**131**:65-86

[12] Wolf SA, Boddeke HW, Kettenmann H. Microglia in physiology and disease. Annual Review of Physiology. 2017;**79**:619-643

[13] Arbor SC, LaFontaine M, Cumbay M. Amyloid-beta Alzheimer targets—Protein processing, lipid rafts, and amyloid-beta pores. The Yale Journal of Biology and Medicine. 2016;**89**(1):5-21

[14] Sarlus H, Heneka MT. Microglia in Alzheimer's disease. The Journal of Clinical Investigation. 2017;**127**(9):3240-3249

[15] Lammich S, Kojro E, Postina R, Gilbert S, Pfeiffer R, Jasionowski M, et al. Constitutive and regulated alpha-secretase cleavage of Alzheimer's amyloid precursor protein by a disintegrin metalloprotease. Proceedings of the National Academy of Sciences of the United States of America. 1999;**96**(7):3922-3927

[16] Fahrenholz F, Gilbert S, Kojro E, Lammich S, Postina R. Alpha-secretase activity of the disintegrin metalloprotease ADAM 10. Influences of domain structure. Annals of the New York Academy of Sciences. 2000;**920**:215-222

[17] Fukumoto H, Cheung BS, Hyman BT, Irizarry MC. Beta-secretase

protein and activity are increased in the neocortex in Alzheimer disease. Archives of Neurology. 2002;**59**(9):1381-1389

[18] Vassar R, Kovacs DM, Yan R, Wong PC. The beta-secretase enzyme BACE in health and Alzheimer's disease: Regulation, cell biology, function, and therapeutic potential. The Journal of Neuroscience. 2009;**29**(41):12787-12794

[19] Nunan J, Williamson NA, Hill AF, Sernee MF, Masters CL, Small DH. Proteasome-mediated degradation of the C-terminus of the Alzheimer's disease beta-amyloid protein precursor: Effect of C-terminal truncation on production of beta-amyloid protein. Journal of Neuroscience Research. 2003;**74**(3):378-385

[20] Luo Y, Bolon B, Kahn S, Bennett BD, Babu-Khan S, Denis P, et al. Mice deficient in BACE1, the Alzheimer's beta-secretase, have normal phenotype and abolished beta-amyloid generation. Nature Neuroscience. 2001;**4**:231-232

[21] Lazarov Lazarov O, Marr RA. Neurogenesis and Alzheimer's disease: At the crossroads. Experimental Neurology. 2010;**223**(2):267-281

[22] Sisodia SS, St George-Hyslop PH. Gamma-secretase, notch, Abeta and Alzheimer's disease: Where do the presenilins fit in? Nature Reviews. Neuroscience. 2002;**3**(4):281-290

[23] Kim BH, Kim JS, Kim JL, Kim YS, Yang TG, Lee MY. Determination of the neutron fluency spectra in the neutron therapy room of KIRAMS. Radiation Protection Dosimetry. 2007;**126**:384-389

[24] Zhang S, Iwata K, Lachenmann MJ, Peng JW, Li S, Stimson ER, et al. The Alzheimer's peptide a beta adopts a collapsed coil structure in water. Journal of Structural Biology. 2000;**130**(2):130-141

[25] Sgourakis NG, Yan Y, McCallum S, Wang C, Garcia AE. The Alzheimer's peptides Aβ40 and 42 adopt distinct conformations in water: A combined MD / NMR study. Journal of Molecular Biology. 2007;**368**(5):1448-1457

[26] Heneka MT, Carson MJ, El Khoury J, Landreth GE, Brosseron F, Feinstein DL, et al. Neuroinflammation in Alzheimer's disease. Lancet Neurology. 2015;**14**(4):388-405

[27] Jack CR Jr, Knopman DS, Jagust WJ, Shaw LM, Aisen PS, Weiner MW, et al. Hypothetical model of dynamic biomarkers of the Alzheimer's pathological cascade. Lancet Neurology. 2010;**9**:119-128

[28] Frost GR, Li YM. The role of astrocytes in amyloid production and Alzheimer's disease. Open Biology. 2017;7:170228

[29] Figley CR, Stroman PW. The role(s) of astrocytes and astrocyte activity in neurometabolism, neurovascular coupling, and the production of functional neuroimaging signals. The European Journal of Neuroscience. 2011;**33**(4):577-588

[30] Aguirre-Rueda D, Guerra-Ojeda S, Aldasoro M, Iradi A, Obrador E, Ortega A, et al. Astrocytes protect neurons from Aβ1-42 peptide-induced neurotoxicity increasing TFAM and PGC-1 and decreasing PPAR-γ and SIRT-1. International Journal of Medical Sciences. 2015;**12**:48-56

[31] Aguirre-Rueda D, Guerra-Ojeda S, Aldasoro M, Iradi A, Obrador E, Mauricio MD, et al. WIN 55,212-2, agonist of cannabinoid receptors, prevents amyloid β1-42 effects on astrocytes in primary culture. PLoS One. 2015b;**10**(4):e0122843

[32] Verkhratsky A, Olabarria M, Noristani HN, Yeh C-Y, Rodriguez JJ.Astrocytes in Alzheimer's

disease. Neurotherapeutics. 2010;7:399-412

[33] Hoffman JM, Dikmen S, Temkin N, Bell KR. Development of posttraumatic stress disorder after mild traumatic brain injury. Archives of Physical Medicine and Rehabilitation. 2012;**93**(2):287-292

[34] Farina N, Page TE, Daley S, Brown A, Bowling A, Basset T, et al. Factors associated with the quality of life of family careers of people with dementia: A systematic review. Alzheimer's & Dementia. 2017;**13**(5):572-581

[35] Markesbery WR. The role of oxidative stress in Alzheimer disease. Archives of Neurology. 1999;**56**:1449-1452

[36] Baggiolini M, Clark-Lewis I. Interleukin-8, a chemotactic and inflammatory cytokine. FEBS Letters. 1992;**307**(1):97-101

[37] McGeer PL, McGeer EG. The inflammatory response system of brain: Implications for therapy of Alzheimer and other neurodegenerative diseases. Brain Research. Brain Research Reviews. 1995;**21**:195-218

[38] Griffin WS, Stanley LC, Ling C, White L, MacLeod V, Perrot LJ, et al. Brain interleukin 1 and S-100 immunoreactivity are elevated in Down syndrome and Alzheimer disease. Proceedings of the National Academy of Sciences of the United States of America. 1989;**86**:7611-7615

[39] Licastro F, Grimaldi LME, Bonafe' M, Martina C, Olivieri F, Cavallone L, et al. Interleukin-6 gene alleles affect the risk of Alzheimer's disease and levels of the cytokine in blood and brain. Neurobiology of Aging. 2003;**24**:921-926

[40] Das S, Potter H. Expression of the Alzheimer amyloid-promoting factor antichymotrypsin is induced in human astrocytes by IL-1. Neuron. 1995;**14**:447-456

[41] Dallagnol KMC, Remor AP, da Silva RA, Prediger RD, Latini A, Aguiar AS Jr. Running for REST: Physical activity attenuates neuroinflammation in the hippocampus of aged mice. Brain, Behavior, and Immunity. 2017;**61**:31-35

[42] Hull M, Strauss S, Berger M, Volk B, Bauer J. The participation of interleukin-6, a stress-inducible cytokine, in the pathogenesis of Alzheimer's disease. Behavioural Brain Research. 1996;**78**(I):37-41

[43] Bhaskar K, Konerth M, Kokiko-Cochran ON, Cardona A, Ransohoff RM, Lamb BT. Regulation of tau pathology by the microglial fractalkine receptor. Neuron. 2010;**68**(1):19-31

[44] Maphis N, Xu G, Kokiko-Cochran ON, Jiang S, Cardona A, Ransohoff RM, et al. Reactive microglia drive tau pathology and contribute to the spreading of pathological tau in the brain. Brain. 2015;**138**:1738-1755

[45] Sanchez-Mejias E, Navarro V, Jimenez S, Sanchez-Mico M, Sanchez-Varo R, Nuñez-Diaz C, et al. Soluble phospho-tau from Alzheimer's disease hippocampus drives microglial degeneration. Acta Neuropathologica. 2016;**132**(6):897-916

[46] Yoshiyama Y, Higuchi M, Zhang B, Huang SM, Iwata N, Saido TC, et al. Synapse loss and microglial activation precede tangles in a P301S tauopathy mouse model. Neuron. 2007;**53**(3):337-351

[47] Asai H, Ikezu S, Tsunoda S, Medalla M, Luebke J, Haydar T, et al. Depletion of microglia and inhibition of exosome synthesis halt tau propagation. Nature Neuroscience. 2015;**18**(11):1584-1593

[48] Gorlovoy P, Larionov S, Pham TT, Neumann H. Accumulation of tau induced in neurites by microglial proinflammatory mediators. The FASEB Journal. 2009;**23**(8):2502-2513

[49] Johnson GV, Stoothoff WH. Tau phosphorylation in neuronal cell function and dysfunction. Journal of Cell Science. 2004;**117**(24):5721-5729

[50] Sato C, Barthélemy NR, Mawuenyega KG, Patterson BW, Gordon BA, Jockel-Balsarotti J, et al. Tau kinetics in neurons and the human central nervous system. Neuron. 2018;**98**(4):861-864

[51] Nizynski B, Dzwolak W, Nieznanski K. Amyloidogenesis of tau protein. Protein Science. 2017;**26**(11):2126-2150

[52] De la De la Fuente-Rocha J. Taupatía en la enfermedad de Alzheimer. Medicina Interna de México. 2017;**33**(4):515-521

[53] Metaxas A, Kempf SJ. Neurofibrillary tangles in Alzheimer's disease: Elucidation of the molecular mechanism by immunohistochemistry and tau protein phospho-proteomics. Neural Regeneration Research. 2016;**11**(10):1579-1581

[54] Ohtsubo K, Izumiyama N, Shimada H, Tachikawa T, Nakamura H. Three-dimensional structure of Alzheimer's neurofibrillary tangles of the aged human brain revealed by the quick-freeze, deep-etch and replica method. Acta Neuropathologica. 1990;**79**(5):480-485

[55] Sheng H, Laskowitz DT, Mackensen GB, Kudo M, Pearlstein RD, Warner DS. Apolipoprotein E deficiency worsens outcome from global cerebral ischemia in the mouse. Stroke. 1999;**30**(5):1118-1124

[56] Jiang T, Zhang YD, Gao Q, Ou Z, Gong PY, Shi JQ, et al. TREM2 ameliorates neuronal tau pathology through suppression of microglial inflammatory response. Inflammation. 2018;**41**(3):811-823

[57] Marin I, Kipnis J. Learning and memory ... and the immune system. Learning & Memory. 2013;**20**(10):601-606

[58] Kitazawa M, Oddo S, Yamasaki TR, Green KN, LaFerla FM. Lipopolysaccharide-induced inflammation exacerbates tau pathology by a cyclin-dependent kinase 5-mediated pathway in a transgenic model of Alzheimer's disease. The Journal of Neuroscience. 2005;**25**(39):8843-8853

[59] Kitazawa M, Trinh DN, LaFerla FM. Inflammation induces tau pathology in inclusion body myositis model via glycogen synthase kinase-3beta. Annals of Neurology. 2008;**64**(1):15-24

[60] Noble W, Garwood C, Stephenson J, Kinsey AM, Hanger DP, Anderton BH. Minocycline reduces the development of abnormal tau species in models of Alzheimer's disease. The FASEB Journal. 2009;**23**(3):739-750

[61] Hunt JB Jr, Nash KR, Placides D, Moran P, Selenica ML, Abuqalbeen F, et al. Sustained arginase 1 expression modulates pathological tau deposits in a mouse model of tauopathy. The Journal of Neuroscience. 2015;**35**(44):14842-14860

[62] Roe AD, Staup MA, Serrats J, Sawchenko PE, Rissman RA. Lipopolysaccharide-induced tau phosphorylation and kinase activity—Modulation, but not mediation, by corticotropin-releasing factor receptors. The European Journal of Neuroscience. 2011;**34**(3):448-456

[63] Harrison JK, Barber CM, Lynch KR. cDNA cloning of a G-protein-coupled receptor expressed in rat spinal cord and

brain related to chemokine receptors. Neuroscience Letters. 1994;**169**:85-89

[64] Gautier EL, Shay T, Miller J, Greter M, Jakubzick C, Ivanov S, et al. Gene-expression profiles and transcriptional regulatory pathways that underlie the identity and diversity of mouse tissue macrophages. Nature Immunology. 2012;**13**:1118-1128

[65] Lee DC, Rizer J, Selenica ML, Reid P, Kraft C, Johnson A, et al. LPS-induced inflammation exacerbates phospho-tau pathology in rTg4510 mice. Journal of Neuroinflammation. 2010;**7**:56

[66] Kitazawa M, Cheng D, Tsukamoto MR, Koike MA, Wes PD, Vasilevko V, et al. Blocking IL-1 signaling rescues cognition, attenuates tau pathology, and restores neuronal beta-catenin pathway function in an Alzheimer's disease model. Journal of Immunology. 2011;**187**(12):6539-6549

[67] Sy M, Kitazawa M, Medeiros R, Whitman L, Cheng D, Lane TE, et al. Inflammation induced by infection potentiates tau pathological features in transgenic mice. The American Journal of Pathology. 2011;**178**(6):2811-2822

[68] Piras A, Collin L, Gruninger F, Graff C, Ronnback A. Autophagic and lysosomal defects in human tauopathies: Analysis of post-mortem brain from patients with familial Alzheimer disease, corticobasal degeneration and progressive supranuclear palsy. Acta Neuropathologica Communications. 2016;**4**:22

[69] Ghosh S, Wu MD, Shaftel SS, Kyrkanides S, LaFerla FM, Olschowka JA, et al. Sustained interleukin-1beta overexpression exacerbates tau pathology despite reduced amyloid burden in an Alzheimer's mouse model. The Journal of Neuroscience. 2013;**33**(11):5053-5064

[70] Mrak RE, Sheng JG, Griffin WS. Correlation of astrocytic S100 beta expression with dystrophic neurites in amyloid plaques of Alzheimer's disease. Journal of Neuropathology and Experimental Neurology. 1996;**55**:273-279

[71] Cho SH, Chen JA, Sayed F, Ward ME, Gao F, Nguyen TA, et al. SIRT1 deficiency in microglia contributes to cognitive decline in aging and neurodegeneration via epigenetic regulation of IL-1beta. The Journal of Neuroscience. 2015;**35**(2):807-818

[72] Liu Z, Condello C, Schain A, Harb R, Grutzendler J. CX3CR1 in microglia regulates brain amyloid deposition through selective protofibrillar amyloidbeta phagocytosis. The Journal of Neuroscience. 2010;**30**(50):17091-17101

[73] Lee S, Xu G, Jay TR, Bhatta S, Kim KW, Jung S, et al. Opposing effects of membrane-anchored CX3CL1 on amyloid and tau pathologies via the p38 MAPK pathway. The Journal of Neuroscience. 2014;**34**(37):12538-12546

[74] Petkau TL, Leavitt BR. Progranulin in neurodegenerative disease. Trends in Neurosciences. 2014;**37**(7):388-398

[75] Takahashi H, Klein ZA, Bhagat SM, Kaufman AC, Kostylev MA, Ikezu T, et al. Opposing effects of progranulin deficiency on amyloid and tau pathologies via microglial TYROBP network. Acta Neuropathologica. 2017;**133**(5):785-807

[76] Nelson PT, Alafuzoff I, Bigio EH, Bouras C, Braak H, Cairns NJ, et al. Correlation of Alzheimer disease neuropathologic changes with cognitive status: A review of the literature. Journal of Neuropathology and Experimental Neurology. 2012;**71**(5):362-381

[77] Zamanian JL, Xu L, Foo LC, Nouri N, Zhou L, Giffard RG, et al.

Genomic analysis of reactive astrogliosis. Journal of Neuroscience. 2012;**32**:6391-6410

[78] Livne-Bar I, Wei J, Liu HH, Alqawlaq S, Won GJ, Tuccitto A, et al. Astrocytederived lipoxins A4 and B4 promote neuroprotection from acute and chronic injury. The Journal of Clinical Investigation. 2017;**127**(12):4403-4414

[79] Vainchtein ID, Chin G, Cho FS, Kelley KW, Miller JG, Chien EC, et al. Astrocyte-derived interleukin-33 promotes microglial synapse engulfment and neural circuit development. Science. 2018;**359**(6381):1269-1273

[80] Sheng JG, Mrak RE, Griffin WST. Microglial interleukin-I-alpha expression in brain regions in Alzheimer's disease—Correlation with neuritic plaque distribution. Neuropathology and Applied Neurobiology. 1995;**21**(4):290-301

[81] Habbas S, Santello M, Becker D, Stubbe H, Zappia G, Liaudet N, et al. Neuroinflammatory TNFalpha impairs memory via astrocyte signaling. Cell. 2015;**163**(7):1730-1741

[82] Anderson MA, Burda JE, Ren Y, Ao Y, O'Shea TM, Kawaguchi R, et al. Astrocyte scar formation aids central nervous system axon regeneration. Nature. 2016;**532**(7598):195-200

[83] Liddelow SA, Barres BA. Reactive astrocytes: Production, function, and therapeutic potential. Immunity. 2017;**46**(6):957-967

[84] Laird FM, Cai H, Savonenko AV, Farah MH, He K, Melnikova T, et al. BACE1, a major determinant of selective vulnerability of the brain to amyloid-b amyloidogenesis, is essential for cognitive, emotional, and synaptic functions. The Journal of Neuroscience. 2005;**25**: 11693-11709

[85] Blasko I, Veerhuis R, Stampfer-Kountchev M, Saurwein-Teissl M, Eikelenboom P, Grubeck-Loebenstein B. Costimulatory effects of interferon-g and interleukin-1b or tumor necrosis factor a on the synthesis of Ab1-40 and Ab1-42 by human astrocytes. Neurobiology of Disease. 2000;**7**:682-689

[86] Jin SM, Cho HJ, Kim YW, Hwang JY, Mook-Jung I. Ab-induced Ca2þ influx regulates astrocytic BACE1 expression via calcineurin/NFAT4 signals. Biochemical and Biophysical Research Communications. 2012;**425**:649-655

[87] Leuba G, Wernli G, Vernay A, Kraftsik R, Mohajeri MH, Saini KD. Neuronal and nonneuronal quantitative BACE immunocytochemical expression in the entorhinohippocampal and frontal regions in Alzheimer's disease. Dementia and Geriatric Cognitive Disorders. 2005;**19**:171-183

[88] Orre M, Kamphuis W, Osborn LM, Jansen AHP, Kooijman L, Bossers K, et al. Isolation of glia from Alzheimer's mice reveals inflammation and dysfunction. Neurobiology of Aging. 2014;**35**:2746-2760

[89] Zhao J, O'Connor T, Vassar R. The contribution of activated astrocytes to Ab production: Implications for Alzheimer's disease pathogenesis. Journal of Neuroinflammation. 2011;**8**:150

[90] Colton CA, Mott RT, Sharpe H, Xu Q, Van Nostrand WE, Vitek MP. Expression profiles for macrophage alternative activation genes in AD and in mouse models of AD. Journal of Neuroinflammation. 2006;**3**:27

[91] Jekabsone A, Mander PK, Tickler A, Sharpe M, Brown GC. Fibrillar beta-amyloid peptide Abeta1-40 activates microglial proliferation via stimulating TNFalpha release and H_2O_2 derived from NADPH oxidase: A cell culture

study. Journal of Neuroinflammation. 2006;**3**:24

[92] Morgan D. Modulation of microglial activation state following passive immunization in amyloid depositing transgenic mice. Neurochemistry International. 2006;**49**:190-194

[93] Malik M, Parikh I, Vasquez JB, Smith C, Tai L, Bu G, et al. Genetics ignite focus on microglial inflammation in Alzheimer's disease. Molecular Neurodegeneration. 2015;**10**:52

[94] Painter MM, Atagi Y, Liu C-C, Rademakers R, Xu H, Fryer JD, et al. TREM2 in CNS homeostasis and neurodegenerative disease. Molecular Neurodegeneration. 2015;**10**:43

[95] Wunderlich P, Glebov K, Kemmerling N, Tien NT, Neumann H, Walter J, et al. Sequential proteolytic processing of the triggering receptor expressed on myeloid cells-2 (TREM2) protein by ectodomain shedding and g-secretase-dependent intramembranous cleavage. The Journal of Biological Chemistry. 2013;**288**:33027-33036

[96] Beach TG, Mcgeer EG. Lamina-specific arrangement of astrocytic gliosis and senile plaques in Alzheimer's disease visual cortex. Brain Research. 1988;**463**:357-361

[97] Sofroniew MV, Vinters HV. Astrocytes: Biology and pathology. Acta Neuropathologica. 2010;**119**:7-35

[98] Huang P, Wang ZY, Tuo Y. The research progression of S100beta as a neurochemistry maker. Fa Yi Xue Za Zhi. 2005;**21**(2):149-151

[99] Daginakatte GC, Gadzinski A, Emnett RJ, Stark JL, Gonzales ER, Yan P, et al. Expression profiling identifies a molecular signature of reactive astrocytes stimulated by cyclic AMP or proinflammatory cytokines. Experimental Neurology. 2008;**210**:261-267

[100] Olabarria M, Noristani HN, Verkhratsky A, Rodríguez JJ. Concomitant astroglial atrophy and astrogliosis in a triple transgenic animal model of Alzheimer's disease. Glia. 2010;**58**:831-838

[101] Sofroniew MV. Molecular dissection of reactive astrogliosis and glial scar formation. Trends in Neurosciences. 2009;**32**:638-647

[102] Bignami A, Eng LF, Dahl D, Uyeda CT. Localization of the glial fibrillary acidic protein in astrocytes by immunofluorescence. Brain Research. 1972;**43**(2):429-435

[103] Carter SF, Schöll M, Almkvist O, Wall A, Engler H, Långström B, et al. Evidence for astrocytosis in prodromal Alzheimer disease provided by ^{11}C-deuterium-L-deprenyl: A multitracer PET paradigm combining ^{11}C-Pittsburgh compound B and ^{18}F-FDG. Journal of Nuclear Medicine. 2012;**53**(1):37-46

[104] Heneka MT, Sastre M, Dumitrescu-Ozimek L, Dewachter I, Walter J, Klockgether T, et al. Focal glial activation coincides with increased BACE1 activation and precedes amyloid plaque deposition in APP[V717I] transgenic mice. Journal of Neuroinflammation. 2005;**2**:22

[105] Jo S, Yarishkin O, Hwang YJ, Chun YE, Park M, Woo DH, et al. GABA from reactive astrocytes impairs memory in mouse models of Alzheimer's disease. Nature Medicine. 2014;**20**(8):886-896

[106] Chun H, An H, Lim J, Woo J, Lee J, Ryu H, et al. Astrocytic proBDNF and tonic GABA distinguish active versus reactive astrocytes in hippocampus. Experimental Neurobiology. 2018;**27**(3):155-170

[107] Oberheim NA, Takano T, Han X, He W, Lin JH, Wang F, et al. Uniquely hominid features of adult human astrocytes. The Journal of Neuroscience. 2009;**29**:3276-3287

[108] Jorda A, Cauli O, Santonja JM, Aldasoro M, Aldasoro C, Obrador E, et al. Changes in chemokines and chemokine receptors expression in a mouse model of Alzheimer's disease. International Journal of Biological Sciences. 2019;**15**:453-463

[109] Valles SL, Dolz-Gaiton P, Gambini J, Borras C, Lloret A, Pallardo FV, et al. Estradiol or genistein prevent Alzheimer's disease-associated inflammation correlating with an increase PPAR gamma expression in cultured astrocytes. Brain Research. 2010;**1312**:138-144

[110] Bardehle S, Krüger M, Buggenthin F, Schwausch J, Ninkovic J, Clevers H, et al. Live imaging of astrocyte responses to acute injury reveals selective juxtavascular proliferation. Nature Neuroscience. 2013;**16**(5):580-586

[111] Constam DB, Philipp J, Malipiero U V, ten Dijke P, Schachner M, Fontana A. Differential expression of transforming growth factor-beta 1, – beta 2, and -beta 3 by glioblastoma cells, astrocytes, and microglia. Journal of Immunology 1992;**148**:1404-1410.

[112] Hu J, Akama KT, Krafft GA, Chromy BA, Van Eldik LJ. Amyloid-b peptide activates cultured astrocytes: Morphological alterations, cytokine induction and nitric oxide release. Brain Research. 1998;**785**:195-206

[113] Johnstone M, Gearing AJ, Miller KM. A central role for astrocytes in the inflammatory response to beta-amyloid; chemokines, cytokines and reactive oxygen species are produced. Journal of Neuroimmunology. 1999;**93**:182-193

[114] Huberman M, Shalit F, Roth-Deri I, Gutman B, Brodie C, Kott E, et al. Correlation of cytokine secretion by mononuclear cells of Alzheimer patients and their disease stage. Journal of Neuroimmunology. 1994;**52**:147-152

[115] De Simone R, Levi G, Aloisi F. Interferon g gene expression in rat central nervous system glial cells. Cytokine. 1998;**10**:418-422

[116] Fultz MJ, Barber SA, Dieffenbach CW, Vogel SN. Induction of IFN-gamma in macrophages by lipopolysaccharide. International Immunology. 1993;**5**:1383-1392

[117] Tarkowski E, Blennow K, Wallin A, Tarkowski A. Intracerebral production of tumor necrosis factor-alpha, a local neuroprotective agent, in Alzheimer disease and vascular dementia. Journal of Clinical Immunology. 1999;**19**:223-230

[118] Tan J, Town T, Crawford F, Mori T, DelleDonne A, Crescentini R, et al. Role of CD40 ligand in amyloidosis in transgenic Alzheimer's mice. Nature Neuroscience. 2002;**5**:1288-1293

[119] Ershler WB, Keller ET. Age-associated increased interleukin-6 gene expression, late-life diseases, and frailty. Annual Review of Medicine. 2000;**51**:245-270

[120] Galimberti D, Venturelli E, Fenoglio C, Guidi I, Villa C, Bergamaschini L, et al. Intrathecal levels of IL-6, IL-11 and LIF in Alzheimer's disease and frontotemporal lobar degeneration. Journal of Neurology. 2008;**255**:539-544

[121] Licastro F, Pedrini S, Caputo L, Annoni G, Davis LJ, Ferri C, et al. Increased plasma levels of interleukin-1, interleukin-6 and a- 1-antichymotrypsin in patients with Alzheimer's disease: Peripheral inflammation or signals from the brain? Journal of Neuroimmunology. 2000;**103**:97-102

[122] Shibata N, Ohnuma T, Takahashi T, Baba H, Ishizuka T, Ohtsuka M, et al. Effect of IL-6 polymorphism on risk of Alzheimer disease: Genotype–phenotype association study in Japanese cases. American Journal of Medical Genetics. 2002;**114**:436-439

[123] Luedecking EK, DeKosky ST, Mehdi H, Ganguli M, Kamboh MI. Analysis of genetic polymorphisms in the transforming growth factorbeta1 gene and the risk of Alzheimer's disease. Human Genetics. 2000;**106**:565-569

[124] Chao CC, Hu S, Frey WH, Ala TA, Tourtellotte WW, Peterson PK, et al. Transforming growth factor beta in Alzheimer's disease. Clinical and Diagnostic Laboratory Immunology. 1994;**1**:109-110

[125] van der Wal EA, Go'mez-Pinilla F, Cotman CW. Transforming growth factor-beta 1 is in plaques in Alzheimer and Down pathologies. Neuroreport. 1993;**4**:69-72

[126] Golde TE, Estus S, Usiak M, Younkin LH, Younkin SG. Expression of beta amyloid protein precursor mRNAs: Recognition of a novel alternatively spliced form and quantitation in Alzheimer's disease using PCR. Neuron. 1990;**4**:253-267

[127] Peress NS, Perillo E. Differential expression of TGF-beta 1, 2 and 3 isotypes in Alzheimer's disease: A comparative immunohistochemical study with cerebral infarction, aged human and mouse control brains. Journal of Neuropathology and Experimental Neurology. 1995;**54**:802-811

[128] Wyss-Coray T, Borrow P, Brooker MJ, Mucke L. Astroglial overproduction of TGF-beta 1 enhances inflammatory central nervous system disease in transgenic mice. Journal of Neuroimmunology. 1997;**77**:45-50

[129] Wyss-Coray T, Lin C, Yan F, Yu G-Q, Rohde M, McConlogue L, et al. TGF-b1 promotes microglial amyloid-b clearance and reduces plaque burden in transgenic mice. Nature Medicine. 2001;7:612-618

[130] LeBlanc AC, Papadopoulos M, Be'lair C, Chu W, Crosato M, Powell J, et al. Processing of amyloid precursor protein in human primary neuron and astrocyte cultures. Journal of Neurochemistry. 1997;**68**:1183-1190

[131] Brugg B, Dubreuil YL, Huber G, Wollman EE, Delhaye-Bouchaud N, Mariani J. Inflammatory processes induce beta-amyloid precursor protein changes in mouse brain. Proceedings of the National Academy of Sciences of the United States of America. 1995;**92**:3032-3035

[132] Goldgaber D, Harris HW, Hla T, Maciag T, Donnelly RJ, Jacobsen JS, et al. Interleukin 1 regulates synthesis of amyloid betaprotein precursor mRNA in human endothelial cells. Proceedings of the National Academy of Sciences of the United States of America. 1989;**86**:7606-7610

[133] Heinrich PC, Castell JV, Andus T. Interleukin-6 and the acute phase response. The Biochemical Journal. 1990;**265**:621-636

[134] Raucher D. Tumor targeting peptides: Novel therapeutic strategies in glioblastoma. Current Opinion in Pharmacology. 2019;**47**:14-19

[135] Friedmann-Morvinski D, Bushong EA, Ke E, Soda Y, Marumoto T, Singer O, et al. Dedifferentiation of neurons and astrocytes by oncogenes can induce gliomas in mice. Science. 2012;**338**(6110):1080-1084

[136] Valles S, Sancho-Tello M, Miñana R, Climent E, Renau-Piqueras J, Guerri C. Glial fibrillary acidic protein expression in rat brain and in radial glia

culture is delayed by prenatal ethanol exposure. Journal of Neurochemistry. 1996;**67**:2425-2433

[137] Sancho-Tello M, Vallés S, Montoliu C, Renau-Piqueras J, Guerri C. Developmental pattern of GFAP and vimentin gene expression in rat brain and in radial glial cultures. Glia. 1995;**15**:157-166

[138] Kuramoto K, Yamamoto M, Suzuki S, Sanomachi T, Togashi K, Seino S, et al. AS602801, an anti-cancer stem cell drug candidate, suppresses gap-junction communication between lung cancer stem cells and astrocytes. Anticancer Research. 2018;**38**:5093-5099

[139] Belanger M, Allaman I, Magistretti PJ. Brain energy metabolism: Focus on astrocyte-neuron metabolic cooperation. Cell Metabolism. 2011;**14**:724-738

[140] Liou GY, Storz P. Reactive oxygen species in cancer. Free Radical Research. 2010;**44**:479-496

[141] Gebicki JM. Oxidative stress, free radicals and protein peroxides. Archives of Biochemistry and Biophysics. 2016;**595**:33-39

[142] Chen Y, Vartiainen NE, Ying W, Chan PH, Koistinaho J, Swanson RA. Astrocytes protect neurons from nitric oxide toxicity by a glutathione-dependent mechanism. Journal of Neurochemistry. 2001;**77**:1601-1610

[143] Fujita T, Tozaki-Saitoh H, Inoue K. P2Y1 receptor signaling enhances neuroprotection by astrocytes against oxidative stress via IL-6 release in hippocampal cultures. Glia. 2009;**57**:244-257

[144] Dringen R, Kussmaul L, Gutterer JM, Hirrlinger J, Hamprecht B. The glutathione system of peroxide detoxification is less efficient in neurons than in astrocytes. Journal of Neurochemistry. 1999;**73**:S106-S106

[145] Huang J, Philbert MA. Distribution of glutathione and glutathione-related enzyme systems in mitochondria and cytosol of cultured cerebellar astrocytes and granule cells. Brain Research. 1995;**680**:16-22

[146] Makar TK, Nedergaard M, Preuss A, Gelbard AS, Perumal AS, Cooper AJ, et al. Ascorbate, glutathione, glutathione disulfide, and enzymes of glutathione metabolism in cultures of chick astrocytes and neurons: Evidence that astrocytes play an important role in antioxidative processes in the brain. Journal of Neurochemistry. 1994;**62**:45-53

[147] Lachapelle MY, Drouin G. Inactivation dates of the human and guinea pig vitamin C genes. Genetica. 2011;**139**:199-207

[148] Covarrubias-Pinto A, Acuña AI, Beltrán FA, Torres-Díaz L, Castro MA. Old things new view: Ascorbic acid protects the brain in neurodegenerative disorders. International Journal of Molecular Sciences. 2015;**16**(12):28194-28217

[149] Castro MA, Beltrán FA, Brauchi S, Concha II. A metabolic switch in brain: Glucose and lactate metabolism modulation by ascorbic acid. Journal of Neurochemistry. 2009;**110**(2):423-440

[150] Persson T, Popescu BO, Cedazo-Minguez A. Oxidative stress in Alzheimer's disease: Why did antioxidant therapy fail? Oxidative Medicine and Cellular Longevity. 2014;**2014**:427318

[151] Allaman I, Gavillet M, Belanger M, Laroche T, Viertl D, Lashuel HA, et al. Amyloid-beta aggregates cause alterations of astrocytic metabolic phenotype: Impact on neuronal

viability. The Journal of Neuroscience. 2010;**30**:3326-3338

[152] Haydon PG. Astrocytes and the modulation of sleep. Current Opinion in Neurobiology. 2017;**44**:28-33

[153] Delekate A, Fuchtemeier M, Schumacher T, Ulbrich C, Foddis M, Petzold GC. Metabotropic P2Y1 receptor signalling mediates astrocytic hyperactivity in vivo in an Alzheimer's disease mouse model. Nature Communications. 2014;**5**:5422

[154] Rossi D, Brambilla L, Valori CF, Crugnola A, Giaccone G, Capobianco R, et al. Defective tumornecrosis factor-alpha-dependent control of astrocyte glutamate release in atransgenic mouse model of Alzheimer disease. The Journal of Biological Chemistry. 2005;**280**:42088-42096

[155] Talantova M, Sanz-Blasco S, Zhang X, Xia P, Akhtar MW, Okamoto S, et al. Abeta induces astrocytic glutamate release, extrasynaptic NMDA receptor activation, and synaptic loss. Proceedings of the National Academy of Sciences of the United States of America. 2013;**110**:E2518-E2527

[156] Valles SL, Borrás C, Gambini J, Furriol J, Ortega A, Sastre J, et al. Oestradiol or genistein rescues neurons from amyloid beta-induced cell death by inhibiting activation of p38. Aging Cell. 2008;**7**:112-118

[157] Liu B, Teschemacher AG, Kasparov S. Neuroprotective potential of astroglia. Journal of Neuroscience Research. 2017;**95**:2126-2139

Chapter 6

Astrocytes: Initiators of and Responders to Inflammation

Allison Soung and Robyn S. Klein

Abstract

We are in the midst of a glial renaissance; astrocytes, essential for brain homeostasis and neuroprotection, have experienced resurgence in focused analyses. New roles in synaptic plasticity, innate immunity and control of recruited immune cells have placed astrocytes at the center of central nervous system functions. Astrocytes have been shown to receive and convey information to all neural cell types in a coordinated effort to respond to injury and infection, initiating reparative mechanisms. Astrocytes detect injury and infection signals from neurons, microglia, oligodendrocytes and endothelial cells, responding by secreting cytokines, chemokines and growth factors, which may activate immune defenses. While regional heterogeneity in astrocyte form and function has been appreciated since the early 1990s, technologic advances have allowed scientists to show only that astrocytes may be as individualized as neurons. Adult astrocytes may undergo a morphological and functional transformation referred to as astrogliosis. Newly generated astrocytes exhibit heterogenous phenotypes; thus, some remove toxic molecules, restore blood-brain barrier function, and promote extracellular matrix components to support axonal growth and repair, while others inhibit neuronal repair and regeneration. This chapter will introduce some of the cellular and molecular components involved in astrocyte responses induced by inflammatory mediators or pathogens during neuroinflammation or neuroinfectious diseases.

Keywords: astrocyte, cytokines, chemokines, pathogens, viruses, bacteria, astrogliosis

1. Introduction

Astrocytes are a principle participant in central nervous system (CNS) responses to neurological disorders or diseases [1–3]. During development and homeostasis, astrocytes coordinate immune responses by regulating microglia activation and blood-brain barrier (BBB) formation [4, 5]. Through dedicated molecular cascades, astrocytes also provide growth factors to neurons, support synapse formation, and help regulate extracellular balance of ions and neurotransmitters, making these glial cells essential for brain homeostasis [6, 7]. In response to CNS injury and disease, astrocytes undergo a process termed astrogliosis, a multifactorial and complex remodeling of astrocytes [7–10]. Despite the use of a single term to describe astrocyte reaction to insult, astrogliosis results in a spectrum of heterogenous changes in a context specific manner that vary with etiology and severity of

IntechOpen

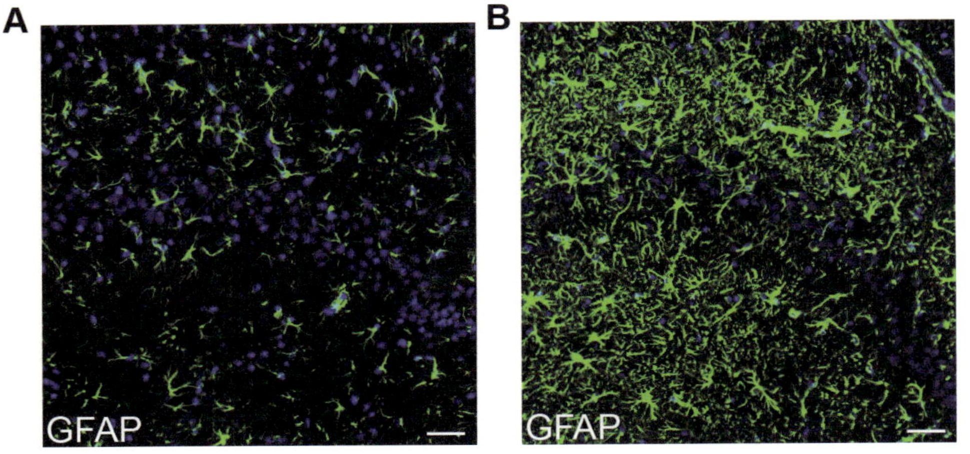

Figure 1.
Astrogliosis is typically characterized by hypertrophy of astrocyte processes. Expression of GFAP+ astrocytes (green) and DAPI (blue) in a mock-infected mouse CA3 hippocampus (A) and in a West Nile virus-infected CA3 hippocampus 7 days post infection (B). Viral infection triggers reactive astrogliosis in the hippocampus, in which the processes of activated astrocytes show hypertrophy.

CNS injury [9–13]. Classically, this process is characterized by upregulation of glial fibrillary acid protein (GFAP) and vimentin, key astrocyte intermediate filaments, and hypertrophy of astrocyte processes [14] (**Figure 1**). Changes in astrocyte biochemistry and physiology that may result in the secretion of anti-inflammatory and pro-inflammatory factors also contribute to this process [10, 15–17].

2. Astrogliosis

Functionally, astrogliosis results in the expression of molecules that provide neurotrophic support to injured neurons, isolate damaged area and CNS inflammation from healthy CNS tissue, rebuild and maintain a compromised BBB, and contribute circuitry remodeling around the lesioned region [7, 9–12, 18]. Consistent with this, studies using animal models of traumatic brain injury, spinal cord injury, and autoimmunity, all reveal that the loss of reactive astrocytes during acute processes leads to the exacerbation of clinical symptoms, recruit of immune molecules, changes in BBB integrity, and neuronal death [7, 10, 19]. The overall goals of these functional reactions are therefore beneficial for the CNS. However, past research has also highlighted detrimental and inhibitory effects of astrogliosis, including augmentation of inflammation, as well as inhibition of neuronal repair and axonal growth [20, 21]. The dual outcomes of astrogliosis highlight the time- and context-specific way this process may be regulated. Future studies of this process may ultimately determine mechanisms to manipulate astrogliosis as a therapeutic target to improve CNS injury outcomes [10, 22].

Astrogliosis is induced and regulated by a variety of extracellular molecules, such as neurotransmitters, steroid hormones, cytokines and neurodegeneration-associated molecules (**Table 1**). Intracellular signaling pathways, such cyclic AMP (cAMP), signal transducer and activator of transcription 3 (STAT3), nuclear factor kappa B (NFκB), Rho-kinase, and calcium have all been observed to induce the expression of GFAP or vimentin [11, 45–47]. Extracellular signaling pathways, including responses to epidermal growth factor (EGF), fibroblast growth factor

Signaling pathway	Injury	Chemokines/cytokines released	Immune/functional outcome	References
ERα signaling	EAE		Reduction of leukocyte molecules	[23]
Gp103/IL-6 signaling	EAE	Downregulation of IL-17 and IFNγ	Reduction of T cell infiltration, inhibition of astrocyte apoptosis, improvement of disease course	[24, 25]
	Infection	Downregulation of IFNγ	Inhibition of astrocyte apoptosis, decrease of pathogen burden	[25]
IL-1β signaling	Traumatic injury, infection		Increase of GFAP expression	[26, 27]
IL-17 signaling	EAE	Upregulation of CXCL2	Increase of leukocyte infiltration, worsen disease course	[28]
IFNγ signaling	EAE	Downregulation of CCL5, IL-1β and TNF	Improved course of disease	[24]
	Traumatic injury		Increase of GFAP expression	[29]
TNFR1 signaling	EAE		Increases of T cell infiltration, worsen disease course	[30]
NFκB signaling	EAE	Upregulation of CCL2, CCL5, CXCL10, IL-1β, IFNγ, and TNF; downregulation of IL-6	Reduction of leukocyte molecules, increase in axon pathology, worsen disease course	[31]
	Ischemia, traumatic injury	Upregulation of CCL2, CCL5, CXCL10, IL-6, TGF-β and TNF	Increase of leukocytes molecules, reduction of GFAP expression, increase of neuronal damage	[31–33]
Notch signaling	Ischemia		Reduction of leukocyte molecules, increase of GFAP expression, increase of astrocyte proliferation	[34, 35]
SHH signaling	EAE		Maintenance of BBB	[36]
Soc3 signaling	Traumatic injury		Increase of leukocyte molecules, increase of GFAP expression	[37]
STAT3 signaling	Traumatic injury		Reduction of leukocyte molecules, inhibition of GFAP expression	[37, 38]
	Traumatic injury		Increase of GFAP and vimentin expression	[39, 40]

Signaling pathway	Injury	Chemokines/cytokines released	Immune/functional outcome	References
TGF-β signaling	Traumatic injury		Increase of leukocyte molecules, increase of GFAP expression, increase of ECM components	[41–43]
	Infection	Inhibition of NFκB signaling; downregulation of CCL5	Reduction of T cell infiltration, decrease of neuronal death	[44]

A variety of intracellular signaling molecules have been shown to induce reactive astrogliosis or to modulate aspects of the reactive astrogliosis process. In response to a range of CNS injuries, all cell types within the CNS, such as neurons, microglia, other astrocytes, endothelium, and pericytes, can release signaling molecules that are able to trigger astrogliosis.
BBB = blood-brain barrier, CCL = chemokine (C-C motif), CXCL = chemokine (C-X-C motif) ligand, ER = estrogen receptor, Gp = glycoprotein, IL = interleukin, IFN = interferon, NFκB = nuclear factor kappa B, EAE = experimental autoimmune encephalomyelitis, ECM = extracellular matrix, SHH = sonic hedgehog, Soc3 = suppressor of cytokine signaling 3, STAT3 = signaling transducer and activator of transcription 3, TGF = transforming growth factor β, TNF = tumor necrosis factor.

Table 1.
Triggers of reactive astrogliosis.

(FGF), sonic hedgehog (SHH), and albumin, can also regulate astrocyte proliferation [9, 48–50]. Specific pro- and anti-inflammatory effects of reactive astrocytes may be regulated separately. Thus, the genetic ablation of STAT3 within astrocytes, or its associated membrane receptor gp130, leads to increased inflammation during autoimmune disease, traumatic injury and infection [24, 37–39, 51], while genetic deletion of NFκB or the suppressor of cytokine signaling 3 (Soc3) signaling pathway in astrocytes decreases the recruitment of immune cells [31, 32, 37]. Furthermore, recruited immune cells release numerous cytokines that may further stimulate astrocyte activation (**Table 1**). In addition, recent studies indicate that microglia critically induce astrogliosis via expression of pro-inflammatory cytokines, including interleukin (IL)-1β, tumor necrosis factor (TNF), and interferon (IFN)-γ [26, 52, 53].

In response to injury, reactive astrocytes were previously believed to migrate to the lesion site. Recent live imaging studies, however, indicate that astrocytes do not migrate towards the lesion site [54]. Instead, astrocytes remain in their tiled-domains and become hypertrophic [54, 55]. Neither proliferation nor migration of astrocytes contribute to the total increase of GFAP positive cells observed at lesion sites. This has led to a new focus on identifying other sources for adult astrocytes. Currently, there is evidence that radial glia, neuronal progenitor cells (NPCs) within the subventricular (SVZ) and subgranular (SGZ) zones, locally proliferating glia, in addition to NG2+ cells may all contribute to newly generated pools of reactive astrocytes after injury [56].

3. Astrocytes as the gatekeeper to the CNS

During homeostasis, astrocyte end-feet enwrap the brain microvascular endothelial cells, helping maintain the integrity of the BBB. Their physical interaction with the BBB allows astrocytes to influence the entry of peripheral immune cells into the CNS during injury or disease as well as modulating their activity once entering the CNS parenchyma. In health, astrocytes, along with multiple other cell types, support the BBB as well as express localizing cues that restrict leukocytes access into the CNS parenchyma [17, 57–59]. However, CNS damage caused by stroke, traumatic injury, infection, autoimmune disease, and neurodegenerative disorders leads to the disruption of the BBB, which may increase the CNS entry of immune cells [24, 25, 38, 39, 60–66].

During injury or infection, astrocytes detect molecular changes in their extracellular environment and in neighboring cells. In stroke, astrocytes become reactive when oxygen and glucose deprivation occurs [67, 68]. In most neurological disorders, the release of neurotransmitters and adenosine triphosphate (ATP) from damaged neurons is detected by astrocytes via P2X and P2Y purinergic receptors [69, 70]. During viral infections, toll-like receptors (TLRs), such as TLR3, 7, and 9, and retinoic acid-inducible gene I (RIG-I)-like receptors (RLRs), are expressed on neurons, astrocytes, and microglia. These receptors are examples of pattern recognition receptors (PPRs) that are differentially activated by pathogen-associated molecular patterns (PAMPs) derived from invading bacteria, fungi, or viruses [71]. Activation of TLRs and RLRs by PAMPs or damage-associated molecular patterns (DAMPs) have been shown to contribute to neuronal damage, induce microglia and astrocyte activation and production of cytokines, including type I IFNs [72–74]. Type I IFNs, along with numerous other innate cytokines, such as IL-6, IL-1β, IFN-γ, and TNF, have been shown to regulate BBB integrity through a variety of different mechanisms that include the regulation of Rho GTPases, activation of matrix metallopeptidase 9 (MMP9), and suppression of other pro-inflammatory cytokines, including IL-1β, IL-6, and TNF [75, 76]. In support of this, the genetic astrocyte-specific deletion of the type I IFN receptor, IFNαβR (IFNAR), results in enhanced BBB permeability in a murine viral infection model [77].

The entry of leukocytes into the CNS parenchyma involves their passage across the BBB, whose permeability is regulated by astrocytes and pericytes, as well as multiple other cell types [57–59]. Once leukocytes traverse the BBB, they localize within perivascular spaces and where they interact with numerous cell types, including astrocytes [78]. Astrocytes, thus, take part in both the recruitment and restriction of leukocytes in the CNS [58, 59, 151]. Their functions, however, occur in a context-specific manner via specific signaling events. It is remarkable how astrocytes are able to respond to a diverse number of signaling mechanisms in the orchestration BBB disruption, the recruitment of leukocytes, and the amplification of their pro-inflammatory effects [17, 57, 79, 80], while also being capable of contributing to BBB repair, restricting leukocyte trafficking, and exerting anti-inflammatory effects that promote the resolution of inflammation [6, 9–11].

4. Reactive astrocytes as a physical barrier

At the site of injury, newly proliferated astrocytes form scars, in which bundles of reactive astrocytes polarize with extracellular matrix (ECM) components and physically surround the lesioned site [38]. The earliest studies focused on the formation of the astrocyte scar and its importance in repairing the BBB after traumatic brain injury [61, 62]. Astrocyte scars form a physical, functional barrier that restricts the entry of leukocytes after traumatic brain injuries, ischemia, neurodegeneration and autoimmune inflammation [37, 38]. This is achieved through the upregulation of ECM proteins, such as fibronectin and laminin, as well as chondroitin sulfate proteoglycans (CSPGs) [41, 42, 81–84]. Structural proteins, such as GFAP and vimentin, have also been shown to be important for the formation of the astrocyte scar [14]. Mice with global genetic deletion of these molecules display increased inflammation and pathology as well as worsened functional outcomes in various CNS injury models, such as ischemia, traumatic injury, autoimmune inflammation, infection, and neurodegeneration [11, 63, 64, 85–88].

The astrocyte scar is also important for localizing immune cells and limiting the invasion of infectious pathogens, to the lesion site. For example, the genetic deletion of GFAP+ cells leads to increases in immune cell infiltrations in murine models of traumatic injury and autoimmunity [60, 61]. Genetic loss of GFAP expression also increases pathogen burden in various infections, including *Staphylococcus aureus* and *Toxoplasma gondii* [89]. Multiple studies have shown that the restriction of leukocyte entry and migration after infection, autoimmune inflammation, and traumatic brain injury is mediated by astrocyte anti-inflammatory functions via the JAK2-STAT3 signaling pathway in GFAP+ cells [25, 38, 39]. The genetic deletion of astrocyte derived STAT3 signaling prevents scar formation and limits immune cell infiltration in a spinal cord injury model [39]. These observations suggest that the astrocyte scar serves as a functional barrier to restrict cytotoxic inflammatory molecules and cells.

Studies genetically deleting essential components of the ECM, such as MMP9, or inhibiting signaling pathways, including Rho/ROCK, to block CSPG activity have shown astrogliosis to exacerbate inflammation after traumatic injury or autoimmune inflammation as well as preventing axonal growth and behavioral recovery [81, 90–92]. The astrocyte scar has also been shown to exhibit a diverse array of molecules known to prevent axonal growth, such as CSPGs, semaphoring 3A, keratan sulfate proteoglycans (KSPGs) and ephrins/Eph receptors [19, 93, 94]. The complexity of astrocytes in producing, recruiting and restricting inflammatory cells and other molecules have made these cells a difficult target for potential therapeutic manipulation.

5. Astrocytes as a regulator of the innate immune response

After CNS injury or infection, reactive astrocytes release molecules that attract, recruit and facilitate the migration of immune cells to the lesion site (**Figure 2**). Astrocytes express leukocyte adhesion molecules, including vascular cell adhesion and intercellular adhesion molecules, in models of ischemia, autoimmunity, and infection [30, 33, 95]. Specifically, in an ischemia model, astrocytes release NF-κB, which increases both vascular cell adhesion and intercellular adhesion molecules [33, 96]. These adhesion molecules promote intercellular interactions that contribute to the trafficking of immune cells to the lesion site.

Like microglia, the resident macrophages of the CNS, astrocytes play a role in innate immune responses by producing cytokines and chemokines, such as type I and II IFNs and TNF, that promote the expression of hundreds of interferon-stimulated genes (ISGs), such as those that participate in inflammatory cell infiltration [97, 98]. Microglia also upregulate the expression of numerous receptors and produce various chemokines after CNS injury, such as chemokine (C-X3-C motif) receptor 1 (CX3CR1) and chemokine (C-C motif) receptor 2 (CCR2) [99]. Similarly, reactive astrocytes also express many of these receptors and chemokines, suggesting that astrocytes and microglia communicate via chemokines. In fact, astrocyte release of chemokines has been shown to be important for attracting peripheral and CNS myeloid cells to the lesion site. In models of traumatic injury and parasitic infection, astrocytes are a source of chemokine (C-C motif) ligand 2 (CCL2) [100, 101]. Astrocytes have also been shown to produce chemokine

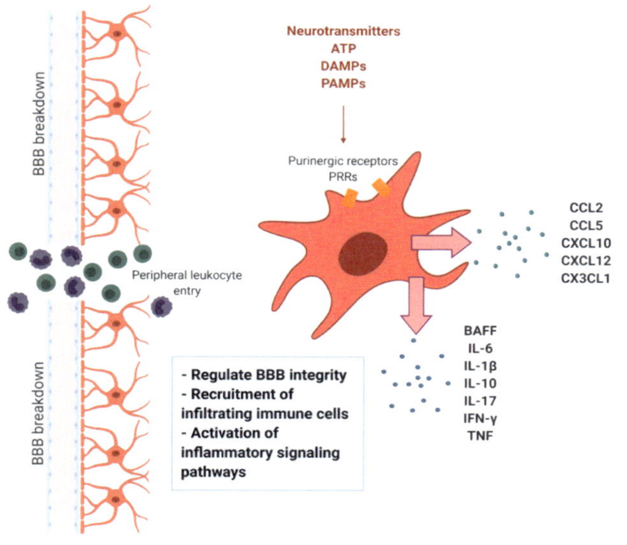

Figure 2.
Reactive astrocytes regulate immune responses. Activation of astrocytes, either directly by purinergic receptors or PRRs or indirectly, promote expression of inflammatory cytokines and chemokines as well as growth factors. The astrocyte response to molecular changes in their microenvironment shapes the recruitment and activation of peripheral immune cells, modulation of the BBB, and cell-cell interaction (not shown). ATP = adenosine triphosphate, BAFF = B-cell activating factor, BBB = blood-brain barrier, CCL = chemokine (C-C motif), CXCL = chemokine (C-X-C motif) ligand, DAMPs = damage-associated molecular pattern, IL = interleukin, IFN = interferon, PAMP = pathogen-associated molecular pattern, PRR = pattern recognition receptor, TNF = tumor necrosis factor.

(C-X3-C motif) ligand 1 (CXC3L1) and CXCL1, detected by monocytes and microglia, in response to viral infection and spinal cord injury, respectively [102–104].

After entry into the brain, or activation within the brain, innate immune cells demonstrate a spectrum of phenotypes, ranging from pro- and anti-inflammatory states, and can express a variety of cytokines and chemokines, including IL-1β, IFN-γ, and TNF, that contribute to neuroinflammation [105]. Reactive astrocytes have a demonstrated role in modulating immune responses by releasing cytokines that stimulate microglia and macrophages to adopt either pro- or anti-inflammatory responses. For example, after injury or infection, astrocytes have been shown to release cytokines, such as IFN-γ, TNF, and IL-12, that shift microglia and macrophages to a more pro-inflammatory phenotype [106, 107]. Under similar conditions, however, astrocytes have also been observed to produced cytokines, including IL-10 and transforming growth factor beta (TGF-β), which can shift monocytes towards a less inflammatory states [108–110]. These findings support the notion that astrocyte responses may be context dependent.

6. Astrocytes as a regulator of the adaptive immune response

During the adaptive immune response, astrocytes are a major source of T and B cells chemoattractants. Reactive astrocytes express CCL5 as well as CXCL10 in infection models, both chemoattractants of T cells [111–113]. In viral infection models, CXCL10 has been shown to be an important ligand for CXCR3 on CD8+ T cells [114]. The recruitment of such CXCR3+ T cells results in improved viral control and survival after infection [115]. In brain samples from patients with multiple sclerosis, astrocytes have been shown to express CXCL12, a T cell chemoattractant, and B-cell activating factor (BAFF), a B cell chemoattractant [116, 117]. Like their influence on microglia and macrophages, cytokines released by reactive astrocytes can shift T cells to adopt either a more beneficial or detrimental phenotype. For example, reactive astrocytes during autoimmunity release pro-inflammatory cytokines, including TNF, IFN-γ, and IL-17, which may induce T cells to adopt a more pro-inflammatory state. However, astrocytes have also been shown to release IL-10 that shifts T cells towards the anti-inflammatory spectrum [23, 24, 32, 118, 119]. Similarly, in a murine spinal cord injury model reactive astrocytes have been shown to release anti-inflammatory TGF-β [31]. Further studies examining the influences of reactive astrocytes on T cells are needed to better understand the long-term effects of astrogliosis on adaptive immune cells during CNS recovery after injury.

7. Reactive astrocytes as a pro-inflammatory regulator

Reactive astrocytes can release a variety of molecular signals that contribute to the inflammatory state of the CNS after injury or disease by directly activating immune defenses with the release of cytokines, chemokines, and other growth factors (**Table 2**). Recent advancements in astrocyte transcriptome analysis have begun to reveal the context specific production of pro-inflammatory molecules by astrocytes as well as molecular triggers that induce their production. Analysis of the astrocyte transcriptome after *in vivo* exposure to lipopolysaccharide (LPS) or infection significantly promoted the production of a pro-inflammatory, neurotoxic molecular profile [26, 52]. However, the astrocyte transcriptome shifts towards an anti-inflammatory, neuroprotective profile in an *in vivo* ischemia model [52]. Future studies should utilize single-cell sequencing techniques to transcriptionally define individual astrocyte responses during health and disease.

Despite the number of astrocyte transcriptome data available, few studies have attempted to elucidate mechanisms and signaling cascades that mediate astrocyte pro-inflammatory production. Recent studies have indicated NFκB and SOC3 as transcriptional regulators of pro-inflammatory astrocytes after a traumatic brain injury and during autoimmune inflammation [31, 32, 37]. In a model of autoimmunity, genetic deletion of astrocyte derived NFκB results in increased expression of ECM components and pro-inflammatory cytokines [129]. Astrocytes have also been shown to release CCL2 and CXCL10 to recruit perivascular leukocytes during autoimmune inflammation [124–126]. While the role of CCL2 and CXCL10 is diverse, evidence suggests that these molecules produced by astrocytes promote leukocyte migration in the CNS parenchyma [124]. In an autoimmune inflammation model, IL-17 inflammatory induction has been shown to be mediated by astrocyte Act1 signaling. Genetically deleting Act1/IL-17 signaling from astrocytes in an EAE model prevents the induction of pro-inflammatory cytokines [28]. Reactive astrocytes can also shift towards a more pro-inflammatory state by overexpressing pro-inflammatory cytokines. In spinal cord injury and autoimmune models, the overexpression of IL-6 in astrocytes leads to increased immune cell infiltration. The proinflammatory cytokine, IL-1β, produced by astrocytes, has also been shown to initiate a signaling cascade that releases vasoactive endothelial growth factor (VEGF), leading to increased BBB permeability and leukocyte leakage [127, 128]. In general, there is also evidence that astrocytes contribute to triggering inflammatory responses due to increases in neuronal activity in epilepsy, neuropathic pain, and stress [130].

Type	Astrocyte molecule	Immune outcome	References
Anti-inflammatory function			
Cytokines	IL-6 and IL-10	Activation of numerous anti-inflammatory signaling pathways	[95, 120]
Growth factors	TGF-β	Activation of numerous anti-inflammatory signaling pathways	[121, 122]
	SHH	Maintenance of BBB, increase of astrocyte proliferation	[36, 123]
Intracellular signaling molecules	STAT3	Suppression of multiple pro-inflammatory signaling pathways	[37–39]
Receptors	ERα and Gp130	Suppression of multiple pro-inflammatory signaling pathways	[23, 25]
Pro-inflammatory function			
Chemokines	CCL2 and CCL5	Increase recruitment of leukocytes	[95, 120, 122, 124, 125]
	CXCL1, CXCL10, and CXCL12	Increase recruitment of leukocytes	[95, 120, 122, 126]
Cytokines	IL-6, IL-17, IL-1β, and TNF	Activation of numerous pro-inflammatory signaling pathways	[95, 120–122]
Growth factors	VEGF	Increase of BBB permeability, increase of leukocyte infiltration	[127, 128]
Intracellular signaling molecules	NFκB and Soc3	Activation of numerous pro-inflammatory signaling pathways	[31, 32, 37]
	Act1	Activation of IL-17-mediated pro-inflammatory signaling pathways	[28]

As immune-competent cells, astrocytes can detect injury signals and respond with the release of cytokine, chemokines, and growth factors as well initiating intracellular signaling pathways. Data suggests that the downstream effects of astrocyte activation is context- and time-dependent, and that both factors as well as other microenvironment stimuli can shift reactive astrocytes to either a more beneficial or detrimental phenotype.
Act = actin, CCL = chemokine (C-C motif), CXCL = chemokine (C-X-C motif) ligand, Gp = glycoprotein, IL = interleukin, IFN = interferon, NFκB = nuclear factor kappa B, EAE = experimental autoimmune encephalomyelitis, SHH = sonic hedgehog, Soc3 = suppressor of cytokine signaling 3, STAT3 = signaling transducer and activator of transcription 3, TGF = transforming growth factor β, TNF = tumor necrosis factor, VEGF = vascular endothelial growth factor.

Table 2.
Pro- and anti-immune responses of astrocyte molecules.

8. Reactive astrocytes as an anti-inflammatory regulator

Despite the growing body of work that suggests pro-inflammatory roles for astrocytes, there is an equal amount of evidence suggesting these cells limit inflammation. Recent loss-of-function experiments have also revealed essential anti-inflammatory roles of astrocytes after a variety of CNS injury and disease states (**Table 2**). These studies have also revealed specific molecular mechanisms that mediate these anti-inflammatory roles. The astrocyte TGF-β response seems to selectively affect astrocyte cytokine and chemokine production after ischemia in murine models. The genetic deletion of TGF-β signaling in astrocytes leads to diffused inflammation and enhances myeloid cell activation [43, 44]. After toxoplasmic encephalitis, the genetic loss of astrocyte TGF-β signaling can lead to the increase of infiltrating T cells. Notably, in both examples, astrocyte TGF-β signaling controls infiltration immune cell number but not necessarily the immune response profile. Astrocyte signaling involving gp130, a receptor for IL-6, or estrogen receptor 1α has also been shown to be anti-inflammatory. In autoimmune and infection models, the genetic deletion of gp130 from astrocytes results in increased inflammatory cytokine production [24, 25]. Similar outcomes, such as increased myeloid infiltration and mortality, are observed in autoimmune models when estrogen receptor 1α is conditional deleted from astrocytes [23]. During autoimmunity, mice deficient in functional IFNγ signaling in astrocytes result in exacerbated disease and mortality due to enhanced leukocyte infiltration and an upregulation of inflammatory gene expression, including CCL1, CCL5, CXC10, and TNF [119]. These mice also had a reduction in anti-inflammatory cytokines, such as IL-10 and IL-27, when compared to mice with functional IFNγ in astrocytes [119].

9. Reactive astrocytes as a neuroprotector of the CNS

In addition to astrocyte regulation of the immune response, these glial cells can respond to CNS injury by altering neuronal function or survival. Neuronal insults result in the release of numerous signals, including increased glutamate production, ATP release and vascular damage. During numerous CNS disease states, including stroke, traumatic injury, epilepsy, neurodegeneration, and viral infection, injured and dying neurons release glutamate, which is harmful to neurons [131–134]. Astrocytes have been shown to take up excessive extracellular glutamate and dampen the neurotransmitter's excitotoxicity on neurons, resulting in decreased neuronal death [135]. *In vitro* studies have also shown that glutamate signaling in astrocytes decrease their production of CCL5, a T cell chemoattractant, reducing overall neuroinflammation [136].

10. Reactive astrocytes as a neurotoxin of the CNS

Inflammation itself can unfortunately impair astrocyte uptake of glutamate, which leads to increased neuronal toxicity and a positive feedback of neuroinflammation [137]; for example, in an *in vitro* study TNF, released by microglia, signals to astrocyte to release glutamate, increasing excitotoxicity [138]. Neuronal injury and death also lead to the release of potassium and ATP. Both potassium and ATP can activate the inflammasome complex, which is an innate immune mechanism that when activated, resulting in the production of proinflammatory cytokines and increased inflammatory responses. The activation of the inflammasome complex, in this case, is through pannexin 1 channels, expressed by astrocytes [139, 140]. Pannexin 1 channels are opened by potassium and ATP, and once opened, activate the inflammasome complex, leading to the increased production of pro-inflammatory mediators, such as IL-1β, reactive oxygen and nitrogen species, and CCL2, a myeloid cell chemoattractant [139, 141–143]. ATP also can induce the release of glutamate from astrocytes, which can contribute to overall excitotoxicity [144]. During health, astrocytes release stored glycogen which is converted to lactate and transported to metabolically support neurons [145]. Neurons can resist excitotoxicity when astrocytes increase their glycogen uptake and lactate delivery [146]. Pro-inflammatory cytokines, including IL-1β as well as IFN-γ, TNF, and IL-6, negatively impacts this process by reducing glycogen storage and lactate transport in astrocytes that is necessary as an energy source of neurons [147, 148].

11. Conclusions

Despite the recent advances in defining the role of astrocytes in regulating neuroinflammation, our understanding of these complex glial cells is only beginning. A few studies have demonstrated astrocyte polarization after various CNS injuries [26, 52, 95]. In this model, "A1" reactive astrocytes are pro-inflammatory, neurotoxic while "A2" reactive astrocytes are anti-inflammatory, neuroprotective. Future research, however, is needed to determine whether, like the inflammatory microglia and macrophages, reactive astrocytes shift phenotypes along a spectrum of responses. The amount of new technology available to researchers will also make it possible to further dissect the complexity of astrocytes. Single-cell transcriptional profiling techniques, specifically, can be used as a tool to identify astrocyte subtypes as well as intracellular signaling networks. This method has already been utilized to reveal distinct astrocyte types with regionally restricted distribution in the healthy mouse brain [149]. A key goal, however, for researchers in the future will be to elucidate signaling networks that are relevant to CNS injury and disease and how immune pathways influence astrocyte reactivity.

While current research focuses primarily on astrocyte interactions with other CNS cell types, such as neurons, microglia, pathogens and infiltrating immune cells, future studies will need to examine how other biologic variables, including age and sex, influence astrocyte effects within the central and peripheral immune systems. Additionally, there is already some evidence that astrocyte immune regulation is influenced by the gut microbiome [150], but the implications and effects of this process on health and disease are unknown.

In summary, astrocytes exhibit diverse and sometimes conflicting roles in the setting of neuroinflammatory diseases. These multipurpose glia cells not only sense and influence damaged neurons but appear to summate multiple signals to develop specific responses that modulate neuroinflammation. It is our hope that understanding how astrocytes receive and response to information as they perform these differential roles will lead to therapies that specifically target astrocytes during CNS injury and disease.

Abbreviations

ATP	adenosine triphosphate
BAFF	B-cell activating factor
BBB	blood-brain barrier
cAMP	cyclic AMP
CCL	chemokine (C-C motif) ligand
CCR	chemokine (C-C motif) receptor
CNS	central nervous system
CSPG	chondroitin sulfate proteoglycan
CXCL	chemokine (C-X-C motif) ligand
CXCR	chemokine (C-X-C motif) receptor
DAMP	damage-associated molecular pattern
ECM	extracellular matrix
EGF	epidermal growth factor
FGF	fibroblast growth factor
GFAP	glial fibrillary acid protein
IFN	interferon
IL	interleukin

ISG	interferon-stimulated gene
KSPG	keratan sulfate proteoglycan
MMP9	matric metallopeptidase 9
NFκB	nuclear factor kappa B
NPC	neural progenitor cell
PAMP	pathogen-associated molecular pattern
PPR	pattern recognition receptor
RIG-I	retinoic acid-inducible gene I
RLRs	RIG-I-like receptor
SGZ	subgranular zone
SHH	sonic hedge hog
SOC3	suppressor of cytokine signaling 3
STAT3	signal transducer and activator of transcription 3
SVZ	subventricular zone
TGF-β	transforming growth factor beta
TLR	toll-like receptor
TNF	tumor necrosis factor
VEGF	vasoactive endothelial growth factor

Author details

Allison Soung[1] and Robyn S. Klein[1,2,3*]

1 Department of Internal Medicine, Washington University School of Medicine, United States

2 Department of Pathology and Immunology, Washington University School of Medicine, United States

3 Department of Neuroscience, Washington University School of Medicine, United States

*Address all correspondence to: rklein@wustl.edu

References

[1] Klein RS, Fricker LD. Heterogeneous expression of car ypeptidase E and proenkephalin mRNAs by cultured astrocytes. Brain Research. 1992;**569**(2):300-310

[2] Bindocci E, Savtchouk I, Liaudet N, Becker D, Carriero G, Volterra A. Neuroscience: Three-dimensional Ca^{2+} imaging advances understanding of astrocyte biology. Science. 2017;**356**(6339):eaai8185

[3] Yoon H, Walters G, Paulsen AR, Scarisbrick IA. Astrocyte heterogeneity across the brain and spinal cord occurs developmentally, in adulthood and in response to demyelination. PLoS One. 2017;**12**(7):e0180697

[4] Bialas AR, Stevens B. TGF-β signaling regulates neuronal C1q expression and developmental synaptic refinement. Nature Neuroscience. 2013;**16**(12):1773-1782

[5] Hurwitz AA, Berman JW, Rashbaum WK, Lyman WD. Human fetal astrocytes induce the expression of blood-brain barrier specific proteins by autologous endothelial cells. Brain Research. 1993;**625**(2):238-243

[6] Barres BA. The mystery and magic of glia: A perspective on their roles in health and disease. Neuron. 2008;**60**(3):430-440

[7] Sofroniew MV, Vinters HV. Astrocytes: Biology and pathology. Acta Neuropathologica. 2010;**119**(1):7-35

[8] Liddelow SA, Barres BA. Reactive astrocytes: Production, function, and therapeutic potential. Immunity. 2017;**46**(6):957-967

[9] Sofroniew MV. Molecular dissection of reactive astrogliosis and glial scar formation. Trends in Neurosciences. 2009;**32**(12):638-647

[10] Sofroniew MV. Astrogliosis. Cold Spring Harbor Perspectives in Biology. 2015;7(2):a020420

[11] Pekny M, Nilsson M. Astrocyte activation and reactive gliosis. Glia. 2005;**50**(4):427-434

[12] Kang W, Hébert JM. Signaling pathways in reactive astrocytes, a genetic perspective. Molecular Neurobiology. 2011;**43**(3):147-154

[13] Anderson MA, Ao Y, Sofroniew MV. Heterogeneity of reactive astrocytes. Neuroscience Letters. 2014;**565**:23-29

[14] Pekny M et al. Abnormal reaction to central nervous system injury in mice lacking glial fibrillary acidic protein and vimentin. The Journal of Cell Biology. 1999;**145**(3):503-514

[15] Sofroniew MV. Multiple roles for astrocytes as effectors of cytokines and inflammatory mediators. The Neuroscientist. 2014;**20**(2):160-172

[16] Pekny M et al. Astrocytes: A central element in neurological diseases. Acta Neuropathologica. 2016;**131**(3):323-345

[17] Burda JE, Sofroniew MV. Reactive gliosis and the multicellular response to CNS damage and disease. Neuron. 2014;**81**(2):229-248

[18] Sofroniew MV. Reactive astrocytes in neural repair and protection. Neuroscience. 2005;**11**(5):400-407

[19] Silver J, Miller JH. Regeneration beyond the glial scar. Nature Reviews. Neuroscience. 2004;**5**(2):146-156

[20] Silver J, Schwab ME, Popovich PG. Central nervous system regenerative failure: Role of oligodendrocytes, astrocytes, and microglia. Cold Spring Harbor Perspectives in Biology. 2015;7(3):a020602

[21] Yiu G, He Z. Glial inhibition of CNS axon regeneration. Nature Reviews Neuroscience. 2006;**7**(8):617-627

[22] Hamby ME, Sofroniew MV. Reactive astrocytes as therapeutic targets for CNS disorders. Neurotherapeutics : The Journal of the American Society for Experimental NeuroTherapeutics. 2010;**7**(4):494-506

[23] Spence RD et al. Neuroprotection mediated through estrogen receptor-α in astrocytes. Proceedings of the National Academy of Sciences of the United States of America. 2011;**108**(21):8867-8872

[24] Haroon F et al. Gp130-dependent astrocytic survival is critical for the control of autoimmune central nervous system inflammation. Journal of Immunology. 2011;**186**(11):6521-6531

[25] Drögemüller K et al. Astrocyte gp130-expression is critical for the control of toxoplasma encephalitis. BMC Proceedings. 2008;**2**(Suppl 1):S10

[26] Garber C, Vasek MJ, Vollmer LL, Sun T, Jiang X, Klein RS. Astrocytes decrease adult neurogenesis during virus-induced memory dysfunction via IL-1 article. Nature Immunology. 2018;**19**(2):151-161

[27] Herx LM, Yong VW. Interleukin-1β is required for the early evolution of reactive astrogliosis following CNS lesion. Journal of Neuropathology and Experimental Neurology. 2001;**60**(10):961-971

[28] Kang Z et al. Astrocyte-restricted ablation of interleukin-17-induced act1-mediated signaling ameliorates autoimmune encephalomyelitis. Immunity. 2010;**32**(3):414-425

[29] Balasingam V, Tejada-Berges T, Wright E, Bouckova R, Yong V. Reactive astrogliosis in the neonatal mouse brain and its modulation by cytokines. The Journal of Neuroscience. 1994;**14**(2):846-856

[30] Gimenez MAT, Sim JE, Russell JH. TNFR1-dependent VCAM-1 expression by astrocytes exposes the CNS to destructive inflammation. Journal of Neuroimmunology. 2004;**151**(1):116-125

[31] Brambilla R et al. Transgenic inhibition of astroglial NF-κB improves functional outcome in experimental autoimmune encephalomyelitis by suppressing chronic central nervous system inflammation. Journal of Immunology. 2009;**182**(5):2628-2640

[32] Brambilla R et al. Inhibition of astroglial nuclear factor κB reduces inflammation and improves functional recovery after spinal cord injury. The Journal of Experimental Medicine. 2005;**202**(1):145-156

[33] Dvoriantchikova G et al. Inactivation of astroglial NF-κB promotes survival of retinal neurons following ischemic injury. The European Journal of Neuroscience. 2009;**30**(2):175-185

[34] Shimada IS, Borders A, Aronshtam A, Spees JL. Proliferating reactive astrocytes are regulated by notch-1 in the peri-infarct area after stroke. Stroke. 2011;**42**(11):3231-3237

[35] Shimada IS, LeComte MD, Granger JC, Quinlan NJ, Spees JL. Self-renewal and differentiation of reactive astrocyte-derived neural stem/progenitor cells isolated from the cortical peri-infarct area after stroke. The Journal of Neuroscience. 2012;**32**(23):7926-7940

[36] Alvarez JI et al. The hedgehog pathway promotes blood-brain barrier integrity and CNS immune quiescence. Science. 2011;**334**(6063):1727-1731

[37] Okada S et al. Conditional ablation of Stat3 or Socs3 discloses a dual role for reactive astrocytes after spinal cord injury. Nature Medicine. 2006;**12**(7):829-834

[38] Wanner IB et al. Glial scar borders are formed by newly proliferated, elongated astrocytes that interact to corral inflammatory and fibrotic cells via STAT3-dependent mechanisms after spinal cord injury. The Journal of Neuroscience. 2013;**33**(31):12870-12886

[39] Herrmann JE et al. STAT3 is a critical regulator of astrogliosis and scar formation after spinal cord injury. The Journal of Neuroscience. 2008;**28**(28):7231-7243

[40] Anderson MA et al. Astrocyte scar formation AIDS central nervous system axon regeneration. Nature. 2016;**532**(7598):195-200

[41] Schachtrup C et al. Fibrinogen triggers astrocyte scar formation by promoting the availability of active TGF-β after vascular damage. The Journal of Neuroscience. 2010;**30**(17):5843-5854

[42] Wang Y, Moges H, Bharucha Y, Symes A. Smad3 null mice display more rapid wound closure and reduced scar formation after a stab wound to the cerebral cortex. Experimental Neurology. 2007;**203**(1):168-184

[43] Cekanaviciute E, Fathali N, Doyle KP, Williams AM, Han J, Buckwalter MS. Astrocytic transforming growth factor-beta signaling reduces subacute neuroinflammation after stroke in mice. Glia. 2014;**62**(8):1227-1240

[44] Cekanaviciute E et al. Astrocytic TGF-β signaling limits inflammation and reduces neuronal damage during central nervous system toxoplasma infection. Journal of Immunology. 2014;**193**(1):139-149

[45] Pekny M, Pekna M. Astrocyte reactivity and reactive astrogliosis: Costs and benefits. Physiological Reviews. 2014;**94**(4):1077-1098

[46] Messing A, Head MW, Galles K, Galbreath EJ, Goldman JE, Brenner M. Fatal encephalopathy with astrocyte inclusions in GFAP transgenic mice. The American Journal of Pathology. 1998;**152**(2):391-398

[47] Gao K et al. Traumatic scratch injury in astrocytes triggers calcium influx to activate the JNK/c-Jun/AP-1 pathway and switch on GFAP expression. Glia. 2013;**61**(12):2063-2077

[48] Gadea A, Schinelli S, Gallo V. Endothelin-1 regulates astrocyte proliferation and reactive gliosis via a JNK/c-Jun signaling pathway. The Journal of Neuroscience. 2008;**28**(10):2394-2408

[49] Levison SW, Jiang FJ, Stoltzfus OK, Ducceschi MH. IL-6-type cytokines enhance epidermal growth factor-stimulated astrocyte proliferation. Glia. 2000;**32**(3):328-337

[50] Sirko S et al. Reactive glia in the injured brain acquire stem cell properties in response to sonic hedgehog. Cell Stem Cell. 2013;**12**(4):426-439

[51] Estrada-Sánchez AM, Rebec GV. Corticostriatal dysfunction and glutamate transporter 1 (GLT1) in Huntington's disease: Interactions between neurons and astrocytes. Basal Ganglia. 2012;**2**(2):57-66

[52] Liddelow SA et al. Neurotoxic reactive astrocytes are induced by activated microglia. Nature. 2017;**541**(7638):481-487

[53] Röhl C, Lucius R, Sievers J. The effect of activated microglia on astrogliosis parameters in astrocyte cultures. Brain Research. 2007;**1129**:43-52

[54] Bardehle S et al. Live imaging of astrocyte responses to acute injury reveals selective juxtavascular proliferation. Nature Neuroscience. 2013;**16**(5):580-586

[55] Wilhelmsson U et al. Redefining the concept of reactive astrocytes as cells that remain within their unique domains upon reaction to injury. Proceedings of the National Academy of Sciences of the United States of America. 2006;**103**(46):17513-17518

[56] Ge W-P, Jia J-M. Local production of astrocytes in the cerebral cortex. Neuroscience. 2016;**323**:3-9

[57] Ransohoff RM, Engelhardt B. The anatomical and cellular basis of immune surveillance in the central nervous system. Nature Reviews. Immunology. 2012;**12**(9):623-635

[58] Owens T, Bechmann I, Engelhardt B. Perivascular spaces and the two steps to neuroinflammation. Journal of Neuropathology and Experimental Neurology. 2008;**67**(12):1113-1121

[59] Wilson EH, Weninger W, Hunter CA. Trafficking of immune cells in the central nervous system. The Journal of Clinical Investigation. 2010;**120**(5):1368-1379

[60] Voskuhl RR et al. Reactive astrocytes form scar-like perivascular barriers to leukocytes during adaptive immune inflammation of the CNS. The Journal of Neuroscience. 2009;**29**(37):11511-11522

[61] Bush TG et al. Leukocyte infiltration, neuronal degeneration, and neurite outgrowth after ablation of scar-forming, reactive astrocytes in adult transgenic mice. Neuron. 1999;**23**(2):297-308

[62] Faulkner JR, Herrmann JE, Woo MJ, Tansey KE, Doan NB, Sofroniew MV. Reactive astrocytes protect tissue and preserve function after spinal cord injury. The Journal of Neuroscience. 2004;**24**(9):2143-2155

[63] Li L et al. Protective role of reactive astrocytes in brain ischemia. Journal of Cerebral Blood Flow and Metabolism. 2008;**28**(3):468-481

[64] Macauley SL, Pekny M, Sands MS. The role of attenuated astrocyte activation in infantile neuronal ceroid lipofuscinosis. The Journal of Neuroscience. 2011;**31**(43):15575-15585

[65] Myer DJ, Gurkoff GG, Lee SM, Hovda DA, Sofroniew MV. Essential protective roles of reactive astrocytes in traumatic brain injury. Brain. 2006;**129**(10):2761-2772

[66] Nawashiro H, Messing A, Azzam N, Brenner M. Mice lacking GFAP are hypersensitive to traumatic cerebrospinal injury. Neuroreport. 1998;**9**(8):1691-1696

[67] Romão LF, De Sousa VO, Neto VM, Gomes FCA. Glutamate activates GFAP gene promoter from cultured astrocytes through TGF-β1 pathways. Journal of Neurochemistry. 2008;**106**(2):746-756

[68] Huang XJ et al. Activation of CysLT receptors induces astrocyte proliferation and death after oxygen-glucose deprivation. Glia. 2008;**56**(1):27-37

[69] Neary JT, Kang Y, Willoughby KA, Ellis EF. Activation of extracellular signal-regulated kinase by stretch-induced injury in astrocytes involves extracellular ATP and P2 purinergic receptors. The Journal of Neuroscience. 2003;**23**(6):2348-2356

[70] Wurm A et al. Purinergic signaling involved in Müller cell function in the mammalian retina. Progress in Retinal and Eye Research. 2011;**30**(5):324-342

[71] Crill EK, Furr-Rogers SR, Marriott I. RIG-I is required for VSV-induced cytokine production by murine glia and acts in combination with DAI to initiate responses to HSV-1. Glia. 2015;**63**(12):2168-2180

[72] Pfefferkorn C et al. Abortively infected astrocytes appear to represent the main source of interferon beta in the virus-infected brain. Journal of Virology. 2016;**90**(4):2031-2038

[73] Bianchi MG, Bardelli D, Chiu M, Bussolati O. Changes in the expression of the glutamate transporter EAAT3/EAAC1 in health and disease. Cellular and Molecular Life Sciences. 2014;**71**(11):2001-2015

[74] Kono H, Rock KL. How dying cells alert the immune system to danger. Nature Reviews. Immunology. 2008;**8**(4):279-289

[75] Daniels BP, Holman DW, Cruz-Orengo L, Jujjavarapu H, Durrant DM, Klein RS. Viral pathogen-associated molecular patterns regulate blood-brain barrier integrity via competing innate cytokine signals. MBio. 2014;**5**(5):e01476-14

[76] Wong D, Dorovini-Zis K, Vincent SR. Cytokines, nitric oxide, and cGMP modulate the permeability of an in vitro model of the human blood-brain barrier. Experimental Neurology. 2004;**190**(2):446-455

[77] Daniels BP et al. Regional astrocyte IFN signaling restricts pathogenesis during neurotropic viral infection. The Journal of Clinical Investigation. 2017;**127**(3):843-856

[78] Brosnan CF, Raine CS. The astrocyte in multiple sclerosis revisited. Glia. 2013;**61**(4):453-465

[79] Goverman J. Autoimmune T cell responses in the central nervous system. Nature Reviews. Immunology. 2009;**9**(6):393-407

[80] Prinz M, Priller J. Microglia and brain macrophages in the molecular age: From origin to neuropsychiatric disease. Nature Reviews. Neuroscience. 2014;**15**(5):300-312

[81] Hsu JYC et al. Matrix metalloproteinase-9 facilitates glial scar formation in the injured spinal cord. The Journal of Neuroscience. 2008;**28**(50):13467-13477

[82] Jones LL, Margolis RU, Tuszynski MH. The chondroitin sulfate proteoglycans neurocan, brevican, phosphacan, and versican are differentially regulated following spinal cord injury. Experimental Neurology. 2003;**182**(2):399-411

[83] Davies SJA, Fitch MT, Memberg SP, Hall AK, Raisman G, Silver J. Regeneration of adult axons in white matter tracts of the central nervous system. Nature. 1997;**390**(6661):680-683

[84] McKeon RJ, Jurynec MJ, Buck CR. The chondroitin sulfate proteoglycans neurocan and phosphacan are expressed by reactive astrocytes in the chronic CNS glial scar. The Journal of Neuroscience. 1999;**19**(24):10778-10788

[85] Middeldorp J, Hol EM. GFAP in health and disease. Progress in Neurobiology. 2011;**93**(3):421-443

[86] Liedtke W, Edelmann W, Chiu FC, Kucherlapati R, Raine CS. Experimental autoimmune encephalomyelitis in mice lacking glial fibrillary acidic protein is characterized by a more severe clinical course and an infiltrative central nervous system lesion. The American Journal of Pathology. 1998;**152**(1):251-259

[87] Liu Z et al. Beneficial effects of gfap/vimentin reactive astrocytes for axonal remodeling and motor behavioral recovery in mice after stroke. Glia. 2014;**62**(12):2022-2033

[88] Kraft AW et al. Attenuating astrocyte activation accelerates plaque pathogenesis in APP/PS1 mice. The FASEB Journal. 2012;**27**(1):187-198

[89] Stenzel W, Soltek S, Schlüter D, Deckert M. The intermediate filament GFAP is important for the control of experimental murine *Staphylococcus aureus*-induced brain abscess and toxoplasma encephalitis. Journal of Neuropathology and Experimental Neurology. 2004;**63**(6):631-640

[90] Monnier PP, Sierra A, Schwab JM, Henke-Fahle S, Mueller BK. The Rho/ROCK pathway mediates neurite growth-inhibitory activity associated with the chondroitin sulfate proteoglycans of the CNS glial scar. Molecular and Cellular Neurosciences. 2003;**22**(3):319-330

[91] Gilbert RJ, McKeon RJ, Darr A, Calabro A, Hascall VC, Bellamkonda RV. CS-4,6 is differentially upregulated in glial scar and is a potent inhibitor of neurite extension. Molecular and Cellular Neurosciences. 2005;**29**(4):545-558

[92] Gates MA, Fillmore H, Steindler DA. Chondroitin sulfate proteoglycan and tenascin in the wounded adult mouse neostriatum *in vitro*: Dopamine neuron attachment and process outgrowth. The Journal of Neuroscience. 1996;**16**(24):8005-8018

[93] Ohtake Y, Li S. Molecular mechanisms of scar-sourced axon growth inhibitors. Brain Research. 2015;**1619**:22-35

[94] Siebert JR, Conta Steencken A, Osterhout DJ. Chondroitin sulfate proteoglycans in the nervous system: Inhibitors to repair. BioMed Research International. 2014;**2014**:845323

[95] Zamanian JL et al. Genomic analysis of reactive astrogliosis. The Journal of Neuroscience. 2012;**32**(18):6391-6410

[96] Ouali Alami N et al. NF-κB activation in astrocytes drives a stage-specific beneficial neuroimmunological response in ALS. The EMBO Journal. 2018;**37**(16):e98697

[97] Cui J, Chen Y, Wang HY, Wang R-F. Mechanisms and pathways of innate immune activation and regulation in health and cancer. Human Vaccines & Immunotherapeutics. 2015;**10**(11):3270-3285

[98] Hu X, Chakravarty SD, Ivashkiv LB. Regulation of interferon and Toll-like receptor signaling during macrophage activation by opposing feedforward and feedback inhibition mechanisms. Immunological Reviews. 2008;**226**:41-56

[99] Cherry JD, Olschowka JA, O'Banion MK. Neuroinflammation and M2 microglia: The good, the bad, and the inflamed. Journal of Neuroinflammation. 2014;**11**

[100] Glabinski AR et al. Chemokine monocyte chemoattractant protein-1 is expressed by astrocytes after mechanical injury to the brain. Journal of Immunology. 1996;**156**(11):4363-4368

[101] Strack A, Asensio VC, Campbell IL, Schlüter D, Deckert M. Chemokines are differentially expressed by astrocytes, microglia and inflammatory leukocytes in toxoplasma encephalitis and critically regulated by interferon-γ. Acta Neuropathologica. 2002;**103**(5):458-468

[102] Pereira CF, Middel J, Jansen G, Verhoef J, Nottet HSLM. Enhanced expression of fractalkine in HIV-1 associated dementia. Journal of Neuroimmunology. 2001;**115**(1):168-175

[103] Pineau I, Sun L, Bastien D, Lacroix S. Astrocytes initiate inflammation in the injured mouse spinal cord by promoting the entry

of neutrophils and inflammatory monocytes in an IL-1 receptor/MyD88-dependent fashion. Brain, Behavior, and Immunity. 2010;**24**(4):540-553

[104] Hughes PM, Botham MS, Frentzel S, Mir A, Perry VH. Expression of fractalkine (CX3CL1) and its receptor, CX3CR1, during acute and chronic inflammation in the rodent CNS. Glia. 2002;**37**:314-327

[105] Kigerl KA, Gensel JC, Ankeny DP, Alexander JK, Donnelly DJ, Popovich PG. Identification of two distinct macrophage subsets with divergent effects causing either neurotoxicity or regeneration in the injured mouse spinal cord. The Journal of Neuroscience. 2009;**29**(43):13435-13444

[106] Fadok VA, Bratton DL, Konowal A, Freed PW, Westcott JY, Henson PM. Macrophages that have ingested apoptotic cells in vitro inhibit proinflammatory cytokine production through autocrine/paracrine mechanisms involving TGF-beta, PGE2, and PAF. The Journal of Clinical Investigation. 1998;**101**(4):890-898

[107] Martinez FO, Helming L, Gordon S. Alternative activation of macrophages: An immunologic functional perspective. Annual Review of Immunology. 2009;**27**(1):451-483

[108] Spera PA, Ellison JA, Feuerstein GZ, Barone FC. IL-10 reduces rat brain injury following focal stroke. Neuroscience Letters. 1998;**251**(3):189-192

[109] Xiaoxing X, E BG, Lijun X, Bing OY, Xinmin X, G GR. Increased brain injury and worsened neurological outcome in interleukin-4 knockout mice after transient focal cerebral ischemia. Stroke. 2011;**42**(7):2026-2032

[110] Ruocco A et al. A transforming growth factor-β antagonist unmasks the neuroprotective role of this endogenous cytokine in excitotoxic and ischemic brain injury. Journal of Cerebral Blood Flow and Metabolism. 1999;**19**(12):1345-1353

[111] Babcock AA, Kuziel WA, Rivest S, Owens T. Chemokine expression by glial cells directs leukocytes to sites of axonal injury in the CNS. The Journal of Neuroscience. 2003;**23**(21):7922-7930

[112] Sauder C et al. Chemokine gene expression in astrocytes of borna disease virus-infected rats and mice in the absence of inflammation. Journal of Virology. 2000;**74**(19):9267-9280

[113] Conant K et al. Induction of monocyte chemoattractant protein-1 in HIV-1 Tat-stimulated astrocytes and elevation in AIDS dementia. Proceedings of the National Academy of Sciences. 1998;**95**(6):3117-3121

[114] Zhang B, Chan YK, Lu B, Diamond MS, Klein RS. CXCR3 mediates region-specific antiviral T cell trafficking within the central nervous system during West Nile virus encephalitis. Journal of Immunology. 2008;**180**(4):2641-2649

[115] Christensen JE et al. Fulminant lymphocytic choriomeningitis virus-induced inflammation of the CNS involves a cytokine-chemokine-cytokine-chemokine cascade. Journal of Immunology. 2009;**182**(2):1079-1087

[116] Krumbholz M et al. Chemokines in multiple sclerosis: CXCL12 and CXCL13 up-regulation is differentially linked to CNS immune cell recruitment. Brain. 2006;**129**(1):200-211

[117] Krumbholz M et al. BAFF is produced by astrocytes and up-regulated in multiple sclerosis lesions and primary central nervous system lymphoma. The Journal of Experimental Medicine. 2005;**201**(2):195-200

[118] Stüve O et al. The role of the MHC class II transactivator in class II expression and antigen presentation by astrocytes and in susceptibility to central nervous system autoimmune disease. Journal of Immunology. 2002;**169**(12):6720-6732

[119] Hindinger C et al. IFN-γ signaling to astrocytes protects from autoimmune mediated neurological disability. PLoS One. 2012;7(7):e42088

[120] Hamby ME, Coppola G, Ao Y, Geschwind DH, Khakh BS, Sofroniew MV. Inflammatory mediators alter the astrocyte transcriptome and calcium signaling elicited by multiple G-protein-coupled receptors. The Journal of Neuroscience. 2012;**32**(42):14489-14510

[121] Meeuwsen S, Persoon-Deen C, Bsibsi M, Ravid R, Van Noort JM. Cytokine, chemokine and growth factor gene profiling of cultured human astrocytes after exposure to proinflammatory stimuli. Glia. 2003;**43**(3):243-253

[122] Jensen CJ, Massie A, De Keyser J. Immune players in the CNS: The astrocyte. Journal of Neuroimmune Pharmacology. 2013;**8**(4):824-839

[123] Pitter KL et al. The SHH/Gli pathway is reactivated in reactive glia and drives proliferation in response to neurodegeneration-induced lesions. Glia. 2014;**62**(10):1595-1607

[124] Kim RY et al. Astrocyte CCL2 sustains immune cell infiltration in chronic experimental autoimmune encephalomyelitis. Journal of Neuroimmunology. 2014;**274**(1):53-61

[125] Moreno M et al. Conditional ablation of astroglial CCL2 suppresses CNS accumulation of M1 macrophages and preserves axons in mice with MOG peptide EAE. The Journal of Neuroscience. 2014;**34**(24):8175-8185

[126] Mills Ko E et al. Deletion of astroglial CXCL10 delays clinical onset but does not affect progressive axon loss in a murine autoimmune multiple sclerosis model. Journal of Neuroinflammation. 2014;**11**(1):105

[127] Argaw AT et al. Astrocyte-derived VEGF-A drives blood-brain barrier disruption in CNS inflammatory disease. The Journal of Clinical Investigation. 2012;**122**(7):2454-2468

[128] Argaw AT, Gurfein BT, Zhang Y, Zameer A, John GR. VEGF-mediated disruption of endothelial CLN-5 promotes blood-brain barrier breakdown. Proceedings of the National Academy of Sciences. 2009;**106**(6):1977-1982

[129] Brambilla R, Morton PD, Ashbaugh JJ, Karmally S, Lambertsen KL, Bethea JR. Astrocytes play a key role in EAE pathophysiology by orchestrating in the CNS the inflammatory response of resident and peripheral immune cells and by suppressing remyelination. Glia. 2014;**62**(3):452-467

[130] Xanthos DN, Sandkühler J. Neurogenic neuroinflammation: Inflammatory CNS reactions in response to neuronal activity. Nature Reviews Neuroscience. 2014;**15**(1):43-53

[131] Kaul M, Garden GA, Lipton SA. Pathways to neuronal injury and apoptosis in HIV-associated dementia. Nature. 2001;**410**(6831):988-994

[132] Maragakis NJ, Rothstein JD. Mechanisms of disease: Astrocytes in neurodegenerative disease. Nature Clinical Practice. Neurology. 2006;**2**(12):679-689

[133] Dong XX, Wang Y, Qin ZH. Molecular mechanisms of excitotoxicity and their relevance to pathogenesis of neurodegenerative

diseases. Acta Pharmacologica Sinica. 2009;**30**(4):379-387

[134] Goodrich GS, Kabakov AY, Hameed MQ, Dhamne SC, Rosenberg PA, Rotenberg A. Ceftriaxone treatment after traumatic brain injury restores expression of the glutamate transporter, GLT-1, reduces regional gliosis, and reduces post-traumatic seizures in the rat. Journal of Neurotrauma. 2013;**30**(16):1434-1441

[135] Lepore AC et al. Reduction in expression of the astrocyte glutamate transporter, GLT1, worsens functional and histological outcomes following traumatic spinal cord injury. Glia. 2011;**59**(12):1996-2005

[136] Besong G et al. Activation of group III metabotropic glutamate receptors inhibits the production of RANTES in glial cell cultures. The Journal of Neuroscience. 2002;**22**(13):5403-5411

[137] Prow NA, Irani DN. The inflammatory cytokine, interleukin-1 beta, mediates loss of astroglial glutamate transport and drives excitotoxic motor neuron injury in the spinal cord during acute viral encephalomyelitis. Journal of Neurochemistry. 2008;**105**(4):1276-1286

[138] Bezzi P et al. CXCR4-activated astrocyte glutamate release via TNFa: Amplification by microglia triggers neurotoxicity. Nature Neuroscience. 2001;**4**(7):702-710

[139] Minkiewicz J, de Rivero Vaccari JP, Keane RW. Human astrocytes express a novel NLRP2 inflammasome. Glia. 2013;**61**(7):1113-1121

[140] Kido Y et al. Regulation of activity of P2X7 receptor by its splice variants in cultured mouse astrocytes. Glia. 2014;**62**(3):440-451

[141] Pelegrin P, Surprenant A. Pannexin-1 mediates large pore formation and interleukin-1beta release by the ATP-gated P2X7 receptor. The EMBO Journal. 2006;**25**(21):5071-5082

[142] Silverman WR et al. The pannexin 1 channel activates the inflammasome in neurons and astrocytes. The Journal of Biological Chemistry. 2009;**284**(27):18143-18151

[143] Xia M, Zhu Y. FOXO3a involvement in the release of TNF-α stimulated by ATP in spinal cord astrocytes. Journal of Molecular Neuroscience. 2013;**51**(3):792-804

[144] Pascual O, Ben Achour S, Rostaing P, Triller A, Bessis A. Microglia activation triggers astrocyte-mediated modulation of excitatory neurotransmission. Proceedings of the National Academy of Sciences. 2012;**109**(4):E197-E205

[145] Pellerin L et al. Evidence supporting the existence of an activity-dependent astrocyte-neuron lactate shuttle. Developmental Neuroscience. 1998;**20**(4-5):291-299

[146] Bliss TM et al. Dual-gene, dual-cell type therapy against an excitotoxic insult by bolstering neuroenergetics. The Journal of Neuroscience. 2004;**24**(27):6202-6208

[147] Bélanger M, Allaman I, Magistretti PJ. Brain energy metabolism: Focus on astrocyte-neuron metabolic cooperation. Cell Metabolism. 2011;**14**(6):724-738

[148] Gavillet M, Allaman I, Magistretti PJ. Modulation of astrocytic metabolic phenotype by proinflammatory cytokines. Glia. 2008;**56**(9):975-989

[149] Zeisel A et al. Molecular architecture of the mouse nervous system resource molecular architecture

of the mouse nervous system. Cell. 2018;**174**:999-1014

[150] Rothhammer V et al. Type i interferons and microbial metabolites of tryptophan modulate astrocyte activity and central nervous system inflammation via the aryl hydrocarbon receptor. Nature Medicine. 2016;**22**(6):586-597

[151] Engelhardt B, Coisne C. Fluids and barriers of the CNS establish immune privilege by confining immune surveillance to a two-walled castle moat surrounding the CNS castle. Fluids and Barriers of the CNS. 2011;**8**(1):4

Chapter 7

Astrocytes in Pathogenesis of Multiple Sclerosis and Potential Translation into Clinic

Izrael Michal, Slutsky Shalom Guy and Revel Michel

Abstract

Astrocytes are the most abundant glial cells in the central nervous system (CNS) and play a pivotal role in CNS homeostasis and functionality. Malfunction of astrocytes was implicated in multiple neurodegenerative diseases, including amyotrophic lateral sclerosis (ALS), Parkinson's disease (PD), Alzheimer's disease (AD), and multiple sclerosis (MS). The involvement of astrocytes in the pathology of neurodegenerative disorders supports the rationale of transplantation of healthy human astrocytes that can potentially compensate for diseased endogenous astrocytes. In this review, we will focus on the roles of astrocytes in the healthy CNS and under MS conditions. We will describe the cell sources and current cell-based therapies for MS with a focus on the potential of astrocyte transplantation. In addition, we will cover immerging early-stage clinical trials in MS that are currently being conducted using cell-based therapies.

Keywords: astrocytes, multiple sclerosis, neurodegenerative diseases, autologous hematopoietic stem cells (AHSC), mesenchymal stem cells (MSC)

1. Multiple sclerosis

Multiple sclerosis (MS) is a chronic, immune-mediated, demyelinating, and degenerative disease of the CNS. The disease leads to permanent neurological disability, including limb weakness, sensory loss, vision disturbances, pain, and muscle spasms [1]. MS is affecting more than 2 million people worldwide, most of them are females between the age of 20 and 40 years. The most prevalent clinical course of the disease (approximately 80% of the cases) is relapsing-remitting MS (RRMS), characterized by a period of functional disability (relapses) and followed by spontaneous improvements (remissions) [1]. With the progression of the disease, most of the patients will develop a course of secondary progressive MS (SPMS), characterized by a steady decline in neurological function, with no phases of remissions [2]. A less common form of MS is primary progressive MS (PPMS), representing approximately 10% of MS cases. PPMS is characterized by a development of gradual progressive disease with no remission phases [2, 3]. Currently, 15 disease-modifying treatments (DMTs) are approved by the FDA for the treatment of MS [4]. The mechanisms of action of these DMTs are diverse; however, they all aim to modulate or suppress the immune system. The current DMTs have benefit in reducing frequency and severeness of relapses and buildup of disability in RRMS; nevertheless, they have only limited impact on the progressive forms of MS [2, 5, 6].

2. Astrocytes in the naive CNS

Although the major players in the onset and development of MS are immune cells, oligodendrocytes, and neurons, astrocytes also play a crucial role in all stages of the pathogenesis of the disease [7]. Astrocytes are the most abundant glial cells in the CNS, making at least 30% of its cell mass in mammalians, having a pivotal role in maintaining the physiologic functions in the CNS [8–10]. Astrocytes can be classified based on their morphological and structural characteristics into two subtypes, namely, protoplasmic and fibrous. Protoplasmic astrocytes are widely distributed in the gray matter, extending processes from their soma to neurons and blood vessels [11]. Their extended end feet are associated with blood vessels to form the glial limiting membrane of the blood-brain barrier (BBB). They also interact with synapses and play an important role in modulation of synaptic functions and uptake of glutamate [12–14]. Conversely, fibrous astrocytes have a starlike appearance, and they are found mainly in the white matter, sending long and thin processes through axonal bundle [15]. Fibrous astrocytes express higher levels of the intermediate filament glial fibrillary acidic protein (GFAP) as compared to protoplasmic astrocyte. Despite the differences in morphology and distribution, both subtypes of astrocytes share many similar functions [16–18].

Astrocytes provide functional support to neurons by maintaining levels of glutamate, extracellular ions, energetic metabolism, pH, and water homeostasis [10, 19]. Astrocytes are also involved in the creation, elimination, and modulation of synapses [20–22]. They modulate the synaptic transmission of neurons by the formation of tripartite synapses that regulate the release of neurotransmitters such as glutamate, d-serine, and gamma-aminobutyric acid (GABA) and by buffering extracellular potassium ions [23–26]. They can also regulate synaptic activity by uptake of neurotransmitters from the synaptic cleft [27, 28]. Astrocytes are important in maintaining the survival of neurons in the CNS, as they secrete neurotrophic and neuroprotective factors such as glial cell line-derived neurotrophic factor (GDNF) and brain-derived neurotrophic factor (BDNF) that directly support neuronal survival [29, 30]. Astrocytes play a pivotal role in formation and maintenance of the blood-brain barrier (BBB), a highly selective physical border that separates the CNS parenchyma from blood circulation through extension of processes of an end-foot membrane that surrounds CNS capillaries [31]. The end-foot membrane contains the channel protein aquaporin-4 (AQP4) and the gap junction protein connexin 43 (Cx43) that allow astrocytes to tightly regulate the selective exchange of water-soluble molecules and ions with blood vessels [32]. In a healthy state, astrocytes constitutively secrete low basal levels of the anti-inflammatory cytokines including transforming growth factor-β (TGF-β) [33] and interleukin-10 (IL-10) [34] to maintain a stable noninflammatory environment. In an inflammatory state, astrocytes change the permeability of the BBB by releasing cytokines such as IL-6, IL-1β, and tumor necrosis factor-α (TNF-α), specifically acting on the endothelial tight junctions of the BBB [35–37]. The close vicinity to blood vessels also allows astrocytes to transfer glucose from the blood to neurons as a source of energy [38]. Astrocytes can also protect neurons from oxidative stress by secretion of antioxidants, such as glutathione and thioredoxin to their coupled neurons [39, 40].

3. Reactive astrocytes

Activation of astrocytes, known also as astrogliosis, is a process that is characterized by proliferation of astrocytes, accompanied by profound morphological and functional changes [41]. Astrocytes become active in response to changes in the CNS homeostasis or under pathological conditions. Cues that lead to astrogliosis include (i) CNS injury that causes the release of damage-associated molecular patterns (DAMPs), (ii) pro-inflammatory cytokines in response to damaged CNS tissue, (iii) pathogen-associated molecular patterns (PAMPs) produced by microbial infection, and (iv) oxidative or chemical stress [42–44]. Although all reactive astrocytes share similar attributes, they can still be distinguished by two different phenotypes, A1 and A2, resembling the M1/M2 states of macrophages [45]. The A1 astrocytes are neurotoxic and induced in response to inflammatory microglia, e.g., those found in neurodegenerative disease such as Huntington's disease (HD) and Parkinson's disease (PD) but also in MS [45, 46]. The A2 reactive astrocytes are formed in response to ischemic damage and, in contrast to the A1-type astrocytes, exhibit anti-inflammatory properties and secrete neurotrophic factors such as BDNF and nerve growth factor (NGF) [93, 45]. Yet, the definition of these two types of reactive astrocytes may be quite elusive, as intermediate phenotypes with mixed characteristics of A1/A2 states were also observed [41]. A1 and A2 astrocytes can appear during different phases of a pathological process and sometimes may even coexist. Their distinct functions allow to attract microglia and T cells by A1 astrocytes at the first stages of the pathology and to support tissue repair by inhibiting inflammation and secreting neurotrophic factors at a later recovery stage [41].

Depending on the severity of the injury, astrogliosis can lead to the formation of a glial scar. The glial scar isolates the inflamed area, restricts the damage to the lesion, and provides structural support to the CNS parenchyma [16]. Based on their environmental cues, reactive astrocytes produce pro- and anti-inflammatory cytokines including IL-1, IL-6, TNF-α, IL-10, and TGF-β [47]. They can also attract circulating leukocytes by secreting chemokines such as CXCL8, CXCL10, CCL2, CCL5, and CCL20 from their end feet at the surface membrane of blood vessels of the BBB [47–49]. Reactive astrocytes also present cell adhesion molecules such as intercellular adhesion molecule 1 (ICAM-1) and vascular cell adhesion protein 1 (VCAM-1), which are important for migration of T cells [50]. Reactive astrocytes can also protect neurons by secretion of neurotrophic factors such as NGF, BDNF, GDNF, and VEGF [51–53]. Although the morphology and activities of reactive astrocytes are well defined, there is no exclusive marker that clearly distinguishes between reactive and nonreactive astrocytes. The major marker of astrogliosis is the intermediate filament GFAP, which is abundant in all astrocyte populations but upregulated upon activation. However, the functional contribution of GFAP in the activation process is still not clear yet [54]. In addition to GFAP, expression of other astrocytic markers is also upregulated in reactive astrocytes, including glutamine synthetase 1, aldehyde dehydrogenase 1 (ALDH1), and S100β [55, 56].

4. Reactive astrocytes in MS

Astrocytes are involved in all stages of the formation and development of the plaques in MS. Their contribution starts already at a very early stage of the lesion, before demyelination is actually seen [57].

Lesions in MS can be classified in four categories.

i. Early pre-active lesions do no not show demyelination damage yet. However, the presence of reactive astrocytes and microglia is the indication for a development of pathological process in the area [58]. Studies in experimental autoimmune encephalomyelitis (EAE) mice suggest that activation of astrocytes can actually occur even before the immune cells cross the BBB into the CNS parenchyma [59].

ii. Active-acute lesions contain hypertrophic astrocytes with enlarged soma and processes comprising high levels of GFAP filaments. In the active-acute plaque, the astrocytes are in close proximity to oligodendrocytes, probably interacting with them. Although the nature of this oligodendrocyte-astrocyte interaction is not completely understood [60–62], it is suggested that astrocytes clear debris of myelin by phagocytosis [63]. Reactive astrocytes in MS may also lose their surface contact with blood vessels of the BBB, enhancing the infiltration of leukocytes to the CNS [57]. The hypertrophic astrocytes also recruit T cells, macrophages, and microglia to the lesion by expressing a set of cell adhesion molecules and chemokines such as ICAM-1 and CCL2 [64–67].

iii. Active-chronic lesions contain a plaque core with a profound active demyelination, which is accompanied by remyelination activity and infiltration of immune cells, especially at the periphery of the lesion. Astrocytes in this type of lesions can be of either A1 or A2 types, and it is suggested that they contribute to the clearance of tissue debris from damaged areas and protect remaining intact regions [45].

In the lesion, reactive astrocytes produce matrix metalloproteinases (MMP), extracellular matrix-remodeling proteins, that changes BBB permeability, allowing immune cell infiltration to the CNS parenchyma and thus inhibiting repair processes [68]. On the other hand, reactive astrocytes also secrete tissue inhibitors of metalloproteinases (TIMPs) in the lesioned area that inhibit the activity of MMPs, help to stabilize BBB permeability, and eventually to promote remyelination [69–71]. Thus, the balance between TIMP and MMP expression can influence the ratio between demyelination and remyelination.

Reactive astrocytes in MS also express a variety of trophic factors that mediate protective and repairing processes in the lesion. Examples of neurotrophic factors which are secreted by astrocytes include neuroprotective factors such as *vascular endothelial growth factor* (VEGF), brain-derived neurotrophic factor (BDNF), neurotrophin-3 (NT3), and insulin-like growth factor-1 (IGF-1) [72, 73]. Reactive astrocytes secrete the cytokine IL-6 that, in addition to its pro-inflammatory activity, also promotes remyelination and neuroprotection [74, 75].

iv. Inactive lesions contain astrocytes with a small cytoplasm and elongated thin processes. The astrocytes in the inactive lesion are rich in GFAP and form a glial scar around the core of the plaque, while occasionally they can be found also within the core [76].

With the progression of MS pathogenesis, reactive hypertrophic astrocytes form a glial scar, which is the most severe grade of astrogliosis, around the core of the demyelinated plaque [10]. The astrocytes in the glial scar form a compact structure that is held by tight junctions on their filament-rich processes [77, 78]. The scar primarily serves as a physical barrier surrounding the demyelinated area, and this prevents widespread of the damage to the surrounding parenchyma [79, 80]. The glial scar also maintains the structure of the BBB, provides structural support, and prevents immune cell infiltration [10, 57]. The glial scar is generally considered as a non-supporting environment for remyelination since it prevents oligodendrocyte progenitor cells (OPCs) from approaching the demyelinated axons surrounded by the glial scar [81, 82].

5. Cell-based therapy

Currently, the available DMTs for MS focus on targeting inflammation processes. These therapies can be divided into two main groups: drugs for the treatment of acute relapses (corticosteroids) [83] and drugs which affect the course of the disease [84]. The second group can be further subdivided into immunosuppressive drugs (e.g., methotrexate and mitoxantrone) and drugs with immunomodulatory activity (e.g., interferon-β [84] and antibodies) [85]. Although these treatments are effective in treating relapsing-remitting MS (RRMS), they show no significant therapeutic benefits in the progressive forms of the disease. A new therapeutic approach with a dual mode of action that is based on tissue repair in addition to immunomodulation has an enormous potential to further attenuate the progression of the disease and to prevent the transition to the progressive course. Cell-based therapies might serve as promising candidates for such a therapy.

The mechanisms of action (MOA) by which therapeutic cells can exert their activities in the CNS include (i) secretion of neurotrophic factors that promote neuronal survival and outgrowth, (ii) reduction of oxidative stress in lesioned areas, (iii) clearance of toxic factors from the CNS environment, (iv) promotion of remyelination, and (v) immunomodulation. In this context, astrocytes hold a promising therapeutic potential, as they share these mechanisms of action [86].

During the last two decades, cell-based therapies from different cell sources were tested in EAE models, and some of them have been further evaluated in clinical trials.

6. Sources of cells for treatment of MS

6.1 Autologous hematopoietic stem cell (AHSC)

Increasing scientific evidence demonstrate that antigen-specific immune response mediates the inflammation process in MS. The immune milieus that depict MS inflammation include (i) immunoglobulins (oligoclonal Igs) that are found in the CSF of the majority of MS patients, but not in their serum [87]; (ii) common clonal T-cell populations in the peripheral blood, cerebrospinal fluid (CSF), and CNS parenchyma [88]; (iii) MHC class II HLA-DRB1 that plays a role in the development of MS [89, 90]; and (iv) specific T-cell receptor (TCR) repertoire in distinct lesions as found in postmortem brains of MS patients [91]. Silent nucleotide exchanges within the V-CDR3-J region of TCR suggest that the corresponding T-cell clones were recruited and stimulated by particular antigens. It was demonstrated that some of the pervasive T-cell clones belonged to the CD8+ compartment, supporting the pathogenic relevance of this T-cell subset [88, 91, 92]. Studies in EAE models and the presence of Th1 and Th17 cells contributed to the notion that self-reactive lymphocytes induce inflammation in response to myelin epitopes [93–95].

One of the approaches to reset the immune system in MS is to use a myeloablative protocol and transplant autologous hematopoietic stem cells (AHSC) similarly to those used in hematologic malignancies [96, 97]. However, immunoablation and reconstitution of the immune system that reset the autoreactive immuno-inflammatory process and restore self-tolerance are still considered as an intensive approach as compared to the current DMTs in MS [97, 98].

6.1.1 Clinical data

One explanation for the therapeutic effect by autologous hematopoietic stem cell transplantation (AHSCT) is reset of the immune system by immune reconstitution following their transplantation. This effect is obtained through deletion of pathogenic clones by a combination of direct ablation and induction of a lymphopenic state. Another explanation might be that the immunosuppression regimen depletes T-cell populations for a long period. AHSCT therapy after immunoablation has been studied for the last 20 years [98]. The results of thousands of patients who have received AHSCT for different types of MS were collected by international transplant registries and showed benefits in a subset of patients with highly active relapsing forms of MS. For instance, recently, a study that was performed in 110 RRMS patients who received AHSCT along with cyclophosphamide (immunosuppressant) and anti-thymocyte globulin, or disease-modifying treatments, was found to prolong the time to disease progression. In the first year, mean of expanded disability status scale (EDSS) scores decreased (improved) from 3.38 to 2.36 in the AHSCT group and increased (worsened) from 3.31 to 3.98 in the DMT group [99]. Other recent trials in MS, mainly in RRMS [99–102], demonstrated a degree of disease stabilization after AHSCT. In addition, recent publications showed a sustained disease attention following AHSCT in a subset of patients with highly active inflammatory disease [103].

The process of immunoablation and reconstitution of the immune system is complicated and includes multiple steps: mobilization of hematopoietic stem cells (HSC), collection and preservation of CD34+ HSCs, immunoablative conditioning, infusion of HSCs, and posttransplant care [104]. It is important to note that immunoablation strategies are also associated with infertility and short-term higher rate of cerebral atrophy that might lead to neurological disability, and hence optimizing treatment regimen is required in order to minimize mortality and morbidity. In addition, this treatment was not found effective for the treatment of primary or secondary progressive MS [105].

6.2 Mesenchymal stem cells

Mesenchymal stem cells (MSCs) are multipotent, non-hematopoietic, stromal cells that can differentiate into mesodermal lineage including osteoblasts, chondrocytes, and adipocytes as well as into ectodermal cells (neurons and glia) and endodermal cells (hepatocytes) [106, 107]. Typically, the bone marrow is used as the source of MSCs. These bone marrow stem cells do not contribute to the formation of blood cells and do not express the hematopoietic stem cell marker CD34 [108]. Alternative tissues that can be used as a source for MSCs include umbilical cord cells that consist of young and most primitive MSCs, adipose tissue, developing tooth bud (molar cells), and the amniotic fluid [109–111]. MSCs present immunomodulatory properties such as activation of regulatory T cells, maturation of dendritic cells, suppression of B- and T-cell proliferation, and inhibition of natural killer functionality. The hypothesis is that the immunomodulatory effect is mediated by paracrine signals and homing of MSCs to the damaged area [112]. Injection of MSCs to EAE animal models demonstrated a slowdown in disease progression, lesser immune cell infiltration, and a decline in demyelination and axonal damage [113, 114]. MSCs were found to possess immunomodulatory effect when administered intraventricular (IVT), intravenously (IV), intrathecally (IT), and intraperitoneally (IP) [113, 115, 116].

6.2.1 Clinical data

In 2007, Mohyeddin et al. were the first to publish their clinical results using MSCs for treatment of MS [117]. The aim of the study was to evaluate the safety and therapeutic potential of autologous MSCs to ameliorate clinical manifestations in MS patients. In this study, 10 MS patients were injected intrathecally with MSCs. The results of the study showed that the use of MSCs is safe, but no significant clinical benefits were observed. In order to provide MSCs with neuromodulatory properties, in addition to their immunomodulatory properties, a few groups differentiated the MSCs into neural-like cells or glial-like cells that secrete neurotrophic factors. IT transplantation of these autologous cells to MS patients demonstrated their safety profile and tolerability [118–120]. Recently, Harris et al. [121] also reported that IT injection of neural-like cells derived from MSCs was safe and well tolerated. The 20 subjects in the clinical trial completed all 60 planned treatments without having serious adverse events. The minor adverse events included transient fever and mild headaches. Posttreatment disability score analysis demonstrated improvement in median EDSS. The beneficial affect was greater in a subset of SPMS patients and in ambulatory subjects (EDSS ≤ 6.5). In addition, 70 and 50% of the subjects demonstrated improved muscle strength and bladder function, respectively [121].

6.3 Neural stem cells and oligodendrocyte precursor cells

In the recent years, clinical trials using cell therapies in MS patients were mainly based on autologous transplantation of MSCs and AHSCs [122]. While showing promising clinical effects, the transplantation of autologous cells is limited to the donor. It would therefore be advantageous to develop allogeneic cell treatments as shelf-products that could be used for large populations of patients. In addition, the potential therapeutic effect of AHSCs and MSCs on MS is mostly mediated through immunomodulatory cues. Finding a cell source that triggers remyelination and tissue, in addition to immunosuppression properties, has a great DMT potential. Neural stem cells (NSCs) can migrate to demyelinated areas and differentiate into neurons and glial-restricted cells (i.e., oligodendrocytes and astrocytes) [7]. NSCs can differentiate to oligodendrocytes that can potentially remyelinate demyelinated axons in MS [123]. The benefits of NSCs might arise not only from their potential to differentiate into oligodendrocytes but also from their capacity to differentiate into astrocytes and neurons, the former having neurotrophic and immunomodulatory properties [86, 123]. Endogenous NSCs are found in germinal niches, such as the subgranular zone (SGZ) of the dentate gyrus and subventricular zone (SVZ) of the lateral ventricles [124, 125]. These NSCs play a pivotal role in early stages of MS, but fail to do so in later stages of the disease. Thus, replenishing endogenous NSCs with allogenic NSCs has a great therapeutic potential. Transplantation of NSCs in EAE animal models demonstrated that the cells can migrate into inflamed white matter plaques and differentiate into oligodendrocytes [126, 127]. Another study showed that transplantation of NSCs derived from induced-pluripotent stem cells (iPSCs) reduced T-cell infiltration as well as white matter damage [128]. To date, no clinical trial in MS evaluated NSCs in MS. A few groups used pluripotent stem cells (human embryonic stem cells or induced-pluripotent stem cells) as a source for neural lineage following an in vitro differentiating protocol [129]. Transplanted hESC-derived NSCs in EAE MS animal models demonstrated neuroprotective and immunosuppressive effect; however, remyelination was not observed [127, 130]. Another study showed that transplantation of iPSC-derived NSCs to

EAE model significantly reduced infiltration of T cells to the lesion and reduced demyelination areas. Consistent with this histopathological improvement, the clinical score of the disease was also rescued in the iPSC-NSC-treated group of mice [128]. Transplantation of hESC-derived OPCs (A2B5$^+$) demonstrated that these cells remyelinate brains of shiverer mice and partially rescue their clinical deficiencies [131–133]. The platelet-derived growth factor α receptor (PDGFAR)-positive OPCs presented even a greater myelinogenic potential [134, 135]. Similarly, intracortical implantation of iPSC-derived OPCs to a nonhuman primate model of progressive multiple sclerosis (MS) showed that the cells can migrate to the lesions and remyelinate denuded axons [136].

6.4 Astrocyte progenitor cells

As discussed above, astrocytes have multiple roles in maintaining the homeostasis of the CNS. Some of the mechanisms of action, which are crucial for the maintenance of the CNS, are postulated to contribute also to the treatment in MS. The diverse modes of action of astrocytes may be more effective in treating MS compared to a single pathway-based drug. Transplantation of healthy astrocytes was proven effective in other neurodegenerative diseases such as ALS [137, 138]. In ALS animal model, it was shown that intrathecal injections of human astrocytes significantly delayed disease onset and improved motor performance compared to sham-injected animals. In this study, the astrocytes were found to secrete various neurotrophic factors and decrease glutamate neurotoxicity [138]. In spinal cord injury (SCI) model, it was demonstrated that transplantation of human astrocytes promotes functional recovery [139–141]. In addition, transplantation of subtype of astroglia was found to possess protective effects against ischemic brain injury [142, 143].

There are several cell sources for human astrocytes. Glial-restricted progenitors (GRPs) represent early cell population of the CNS that can self-renew and give rise to astrocytes and oligodendrocytes. GRPs can be isolated from human fetal tissues [144]. In vivo transplantation of human GRPs into the spinal cord-injured animals showed that the cells can survive and differentiate into astrocytes [139, 140]. However, human astrocytes from primary brain tissue, obtained from cadaveric donors, are challenging due to limited availability and robustness.

Other sources for derivation of astrocytes include pluripotent stem cells (PSC) such as embryonic stem cells and induced-pluripotent stem cells (iPSCs) [145]. These sources potentially provide unlimited supply of cells for clinical use. Methods for producing neural precursor cells from PSCs and their further differentiation into glial lineage were demonstrated in pioneering studies in animal models of neurodevelopment. In these studies, the key steps for neural commitment in vivo were identified and recapitulated in a stepwise process in culture. Specific commitment of pluripotent stem cells toward astrocytes can be achieved using factors such as sonic hedgehog (SHH), Wnt proteins, fibroblast growth factors (FGFs), epidermal growth factors (EGFs), retinoic acid (RA), and bone morphogenetic protein (BMP) [146–150]. Most recently, direct-reprogramming approaches of somatic cells into neural cells and astrocytes, including transduction of specified transcription factors or by using a combination of defined chemical, have been reported [151]. Caiazzo et al. [152] described a conversion of mouse fibroblast into astrocytes (iAstrocytes), which are comparable to endogenous astrocytes. This was carried out by transducing the transcription factors *nuclear factors* IA and IB (NFIA, NFIB) and SOX9. Another approach for direct conversion or reprogramming of mammalian fibroblasts into astrocytes is by culturing the cells in the presence of a cocktail of small molecules that includes histone deacetylase inhibitor VPA, TGFβ, and GSK3β inhibitor CHIR99021, among other factors [153].

7. Conclusions

MS is a multifactorial disease involving dysregulation of molecular pathways and immunomodulatory processes. Transplantation of healthy functional cells that can affect the CNS via diverse mechanisms of action that work in parallel such as anti-inflammatory, immunomodulatory, clearance of the toxic environment, secretion of neurotrophic factors, and triggering remyelination has great therapeutic potential in treating multiple sclerosis. Yet, bringing new cell-based therapies to the clinic faces a few challenges, e.g., what is the optimal injection site in the CNS, and what cell dose will be effective? In MS the demyelinated lesions are spread throughout the CNS, and it still not clear whether the transplanted cells have long-distance migratory capacity to reach these plaques from their injection site. Once the cells reach to lesion, it is still questionable whether they can remyelinate axons under a hostile inflammatory environment. Finally, the safety profile of transplanted cells and their long-term tumorigenic potential should be further tested.

Abbreviations

AD	Alzheimer's disease
AHSC	autologous hematopoietic stem cell
ALDH1	aldehyde dehydrogenase
ALS	amyotrophic lateral sclerosis
AQP4	aquaporin-4
BBB	blood-brain barrier
BDNF	brain-derived neurotrophic factor
BMP	bone morphogenetic protein
CCL2	chemokine C-C motif ligand
CNS	central nervous system
CSF	cerebrospinal fluid
Cx43	connexin 43
CXCL	chemokine C-X-C motif ligand
DAMP	damage-associated molecular pattern
DMT	disease-modifying treatment
EAE	experimental autoimmune encephalomyelitis
EDSS	expanded disability status scale
EGF	epidermal growth factor
FGF	fibroblast growth factor
GABA	gamma-aminobutyric acid
GDNF	glial cell line-derived neurotrophic factor
GFAP	glial fibrillary acidic protein
HD	Huntington's disease
HSC	hematopoietic stem cells
ICAM-1	intercellular adhesion molecule 1
IGF-1	insulin-like growth factor-1
IL	interleukin
IP	intraperitoneally
iPSC	induced-pluripotent stem cell
IT	intrathecally
IV	intravenously
IVT	intraventricular
MMP	matrix metalloproteinases
MOA	mechanisms of action

MS	multiple sclerosis
MSC	mesenchymal stem cells
NFI	nuclear factor I
NGF	nerve growth factor
NSC	neural stem cell
NT-3	neurotrophin-3
PAMP	pathogen-associated molecular pattern
PD	Parkinson's disease
PPMS	primary progressive multiple sclerosis
PSC	pluripotent stem cell
RA	retinoic acid
RRMS	relapsing-remitting multiple sclerosis
SCI	spinal cord injury
SGZ	sub granular zone
SHH	sonic hedgehog
SPMS	secondary progressive multiple sclerosis
SVZ	subventricular zone
TCR	T-cell receptor
TGF-β	transforming growth factor-β
TIMP	tissue inhibitors of metalloproteinases
TNF-α	tumor necrosis factor-α
VCAM-1	vascular cell adhesion protein 1
VEGF	vascular endothelial growth factor

Author details

Izrael Michal[1*], Slutsky Shalom Guy[1] and Revel Michel[1,2]

1 Department of Neurodegenerative Diseases at Kadimastem Ltd., Ness Ziona, Israel

2 Department of Molecular Genetics, Weizmann Institute of Science, Rehovot, Israel

*Address all correspondence to: m.izrael@kadimastem.com

References

[1] Kobelt G et al. New insights into the burden and costs of multiple sclerosis in Europe. Multiple Sclerosis. 2017;**23**(8):1123-1136

[2] Lublin FD et al. Defining the clinical course of multiple sclerosis: The 2013 revisions. Neurology. 2014;**83**(3):278-286

[3] Polman CH et al. Diagnostic criteria for multiple sclerosis: 2010 revisions to the McDonald criteria. Annals of Neurology. 2011;**69**(2):292-302

[4] Cree BAC, Mares J, Hartung HP. Current therapeutic landscape in multiple sclerosis: An evolving treatment paradigm. Current Opinion in Neurology. 2019;**32**(3):365-377

[5] Wingerchuk DM, Carter JL. Multiple sclerosis: Current and emerging disease-modifying therapies and treatment strategies. Mayo Clinic Proceedings. 2014;**89**(2):225-240

[6] Kutzelnigg A, Lassmann H. Pathology of multiple sclerosis and related inflammatory demyelinating diseases. Handbook of Clinical Neurology. 2014;**122**:15-58

[7] Fitzner D, Simons M. Chronic progressive multiple sclerosis: Pathogenesis of neurodegeneration and therapeutic strategies. Current Neuropharmacology. 2010;**8**(3):305-315

[8] Lundgaard I et al. White matter astrocytes in health and disease. Neuroscience. 2014;**276**:161-173

[9] Barres BA. The mystery and magic of glia: A perspective on their roles in health and disease. Neuron. 2008;**60**(3):430-440

[10] Sofroniew MV, Vinters HV. Astrocytes: Biology and pathology. Acta Neuropathologica. 2010;**119**(1):7-35

[11] Nedergaard M, Ransom B, Goldman SA. New roles for astrocytes: Redefining the functional architecture of the brain. Trends in Neurosciences. 2003;**26**(10):523-530

[12] Oliet SH, Piet R, Poulain DA. Control of glutamate clearance and synaptic efficacy by glial coverage of neurons. Science. 2001;**292**(5518):923-926

[13] Rothstein JD et al. Knockout of glutamate transporters reveals a major role for astroglial transport in excitotoxicity and clearance of glutamate. Neuron. 1996;**16**(3):675-686

[14] Henneberger C et al. Long-term potentiation depends on release of D-serine from astrocytes. Nature. 2010;**463**(7278):232-236

[15] Butt AM, Duncan A, Berry M. Astrocyte associations with nodes of Ranvier: Ultrastructural analysis of HRP-filled astrocytes in the mouse optic nerve. Journal of Neurocytology. 1994;**23**(8):486-499

[16] McKinnon PJ, Margolskee RF. SC1: A marker for astrocytes in the adult rodent brain is upregulated during reactive astrocytosis. Brain Research. 1996;**709**(1):27-36

[17] Brown AM, Ransom BR. Astrocyte glycogen and brain energy metabolism. Glia. 2007;**55**(12):1263-1271

[18] Marin-Padilla M. Prenatal development of fibrous (white matter), protoplasmic (gray matter), and layer I astrocytes in the human cerebral cortex: A Golgi study. The Journal of Comparative Neurology. 1995;**357**(4):554-572

[19] Browne EC, Abbott BM. Recent progress towards an effective treatment of amyotrophic lateral sclerosis using the SOD1 mouse model in a preclinical setting. European Journal of Medicinal Chemistry. 2016;**121**:918-925

[20] Fields RD. Oligodendrocytes changing the rules: Action potentials in glia and oligodendrocytes controlling action potentials. The Neuroscientist. 2008;**14**(6):540-543

[21] Hong S et al. Complement and microglia mediate early synapse loss in Alzheimer mouse models. Science. 2016;**352**(6286):712-716

[22] Adamsky A et al. Astrocytic activation generates de novo neuronal potentiation and memory enhancement. Cell. 2018;**174**(1):59-71 e14

[23] Bezzi P, Volterra A. A neuron-glia signalling network in the active brain. Current Opinion in Neurobiology. 2001;**11**(3):387-394

[24] Ventura R, Harris KM. Three-dimensional relationships between hippocampal synapses and astrocytes. The Journal of Neuroscience. 1999;**19**(16):6897-6906

[25] Santello M, Volterra A. Neuroscience: Astrocytes as aide-memoires. Nature. 2010;**463**(7278):169-170

[26] Sun W et al. Glutamate-dependent neuroglial calcium signaling differs between young and adult brain. Science. 2013;**339**(6116):197-200

[27] Kang J et al. Astrocyte-mediated potentiation of inhibitory synaptic transmission. Nature Neuroscience. 1998;**1**(8):683-692

[28] Haydon PG. Neuroglial networks: Neurons and glia talk to each other. Current Biology. 2000;**10**(19):R712-R714

[29] Ludwin SK et al. Astrocytes in multiple sclerosis. Multiple Sclerosis. 2016;**22**(9):1114-1124

[30] Magistretti PJ. Neuron-glia metabolic coupling and plasticity. The Journal of Experimental Biology. 2006;**209**(Pt 12):2304-2311

[31] Iliff JJ et al. A paravascular pathway facilitates CSF flow through the brain parenchyma and the clearance of interstitial solutes, including amyloid beta. Science Translational Medicine. 2012;**4**(147):147ra111

[32] Brambilla R. The contribution of astrocytes to the neuroinflammatory response in multiple sclerosis and experimental autoimmune encephalomyelitis. Acta Neuropathologica. 2019;**137**(5):757-783

[33] John GR, Lee SC, Brosnan CF. Cytokines: Powerful regulators of glial cell activation. The Neuroscientist. 2003;**9**(1):10-22

[34] Cannella B, Raine CS. The adhesion molecule and cytokine profile of multiple sclerosis lesions. Annals of Neurology. 1995;**37**(4):424-435

[35] Kim JH et al. Blood-neural barrier: Intercellular communication at glio-vascular interface. Journal of Biochemistry and Molecular Biology. 2006;**39**(4):339-345

[36] Didier N et al. Secretion of interleukin-1beta by astrocytes mediates endothelin-1 and tumour necrosis factor-alpha effects on human brain microvascular endothelial cell permeability. Journal of Neurochemistry. 2003;**86**(1):246-254

[37] Schwaninger M et al. Bradykinin induces interleukin-6 expression in astrocytes through activation of nuclear factor-kappa B. Journal of Neurochemistry. 1999;**73**(4):1461-1466

[38] Villarroya H et al. Myelin-induced experimental allergic encephalomyelitis in Lewis rats: Tumor necrosis factor alpha levels in serum and cerebrospinal fluid immunohistochemical expression in glial cells and macrophages of optic nerve and spinal cord. Journal of Neuroimmunology. 1996;**64**(1):55-61

[39] Masutani H et al. Thioredoxin as a neurotrophic cofactor and an important regulator of neuroprotection. Molecular Neurobiology. 2004;**29**(3):229-242

[40] Anderson CM, Bergher JP, Swanson RA. ATP-induced ATP release from astrocytes. Journal of Neurochemistry. 2004;**88**(1):246-256

[41] Pekny M et al. Astrocytes: A central element in neurological diseases. Acta Neuropathologica. 2016;**131**(3):323-345

[42] Rothhammer V, Quintana FJ. Control of autoimmune CNS inflammation by astrocytes. Seminars in Immunopathology. 2015;**37**(6):625-638

[43] Jensen CJ, Massie A, De Keyser J. Immune players in the CNS: The astrocyte. Journal of Neuroimmune Pharmacology. 2013;**8**(4):824-839

[44] Farina C, Aloisi F, Meinl E. Astrocytes are active players in cerebral innate immunity. Trends in Immunology. 2007;**28**(3):138-145

[45] Liddelow SA, Barres BA. Reactive astrocytes: Production, function, and therapeutic potential. Immunity. 2017;**46**(6):957-967

[46] Liddelow SA et al. Neurotoxic reactive astrocytes are induced by activated microglia. Nature. 2017;**541**(7638):481-487

[47] Dong Y, Benveniste EN. Immune function of astrocytes. Glia. 2001;**36**(2):180-190

[48] Williams WM et al. Do beta-defensins and other antimicrobial peptides play a role in neuroimmune function and neurodegeneration? Scientific World Journal. 2012;**2012**:905785

[49] Mahida YR, Cunliffe RN. Defensins and mucosal protection. Novartis Foundation Symposium. 2004;**263**:71-77; discussion 77-84, 211-218

[50] Lee SJ, Benveniste EN. Adhesion molecule expression and regulation on cells of the central nervous system. Journal of Neuroimmunology. 1999;**98**(2):77-88

[51] Sofroniew MV. Molecular dissection of reactive astrogliosis and glial scar formation. Trends in Neurosciences. 2009;**32**(12):638-647

[52] Gimenez MA, Sim JE, Russell JH. TNFR1-dependent VCAM-1 expression by astrocytes exposes the CNS to destructive inflammation. Journal of Neuroimmunology. 2004;**151**(1-2):116-125

[53] Ransohoff RM, Kivisakk P, Kidd G. Three or more routes for leukocyte migration into the central nervous system. Nature Reviews. Immunology. 2003;**3**(7):569-581

[54] Korfias S et al. Serum S-100B protein as a biochemical marker of brain injury: A review of current concepts. Current Medicinal Chemistry. 2006;**13**(30):3719-3731

[55] Khakh BS, Sofroniew MV. Diversity of astrocyte functions and phenotypes in neural circuits. Nature Neuroscience. 2015;**18**(7):942-952

[56] Yang Y et al. Molecular comparison of GLT1+ and ALDH1L1+ astrocytes *in vivo* in astroglial reporter mice. Glia. 2011;**59**(2):200-207

[57] Brosnan CF, Raine CS. The astrocyte in multiple sclerosis revisited. Glia. 2013;**61**(4):453-465

[58] Alvarez JI et al. Focal disturbances in the blood-brain barrier are associated with formation of neuroinflammatory lesions. Neurobiology of Disease. 2015;**74**:14-24

[59] Pham H et al. The astrocytic response in early experimental autoimmune encephalomyelitis occurs across both the grey and white matter compartments. Journal of Neuroimmunology. 2009;**208**(1-2):30-39

[60] Victor J et al. Pharmacokinetics and efficacy of delayed-action hydroquinidine in patients with ventricular extrasystole. Annales de cardiologie et d'angéiologie (Paris). 1987;**36**(3):163-166

[61] Popiela A, Zdrojewicz Z, Kasiak J. Clinical problems related to postmenopausal osteoporosis. Polski Tygodnik Lekarski. 1991;**46**(30-31):568-570

[62] Ludwin SK. An autoradiographic study of cellular proliferation in remyelination of the central nervous system. The American Journal of Pathology. 1979;**95**(3):683-696

[63] Skripuletz T et al. Astrocytes regulate myelin clearance through recruitment of microglia during cuprizone-induced demyelination. Brain. 2013;**136**(Pt 1):147-167

[64] Bullard DC et al. Intercellular adhesion molecule-1 expression is required on multiple cell types for the development of experimental autoimmune encephalomyelitis. Journal of Immunology. 2007;**178**(2):851-857

[65] Sorensen TL et al. Expression of specific chemokines and chemokine receptors in the central nervous system of multiple sclerosis patients. The Journal of Clinical Investigation. 1999;**103**(6):807-815

[66] Ponath G et al. Myelin phagocytosis by astrocytes after myelin damage promotes lesion pathology. Brain. 2017;**140**(2):399-413

[67] Kim RY et al. Astrocyte CCL2 sustains immune cell infiltration in chronic experimental autoimmune encephalomyelitis. Journal of Neuroimmunology. 2014;**274**(1-2):53-61

[68] Gardner J, Ghorpade A. Tissue inhibitor of metalloproteinase (TIMP)-1: The TIMPed balance of matrix metalloproteinases in the central nervous system. Journal of Neuroscience Research. 2003;**74**(6):801-806

[69] Moore CS et al. How factors secreted from astrocytes impact myelin repair. Journal of Neuroscience Research. 2011;**89**(1):13-21

[70] Moore CS et al. Astrocytic tissue inhibitor of metalloproteinase-1 (TIMP-1) promotes oligodendrocyte differentiation and enhances CNS myelination. The Journal of Neuroscience. 2011;**31**(16):6247-6254

[71] Pagenstecher A et al. Differential expression of matrix metalloproteinase and tissue inhibitor of matrix metalloproteinase genes in the mouse central nervous system in normal and inflammatory states. The American Journal of Pathology. 1998;**152**(3):729-741

[72] Ludwin SK. The pathogenesis of multiple sclerosis: Relating human pathology to experimental studies. Journal of Neuropathology and Experimental Neurology. 2006;**65**(4):305-318

[73] Nair A, Frederick TJ, Miller SD. Astrocytes in multiple sclerosis: A product of their environment.

Cellular and Molecular Life Sciences. 2008;**65**(17):2702-2720

[74] Willenborg DO et al. Cytokines and murine autoimmune encephalomyelitis: Inhibition or enhancement of disease with antibodies to select cytokines, or by delivery of exogenous cytokines using a recombinant vaccinia virus system. Scandinavian Journal of Immunology. 1995;**41**(1):31-41

[75] Rodriguez M et al. Recombinant human IL-6 suppresses demyelination in a viral model of multiple sclerosis. Journal of Immunology. 1994;**153**(8):3811-3821

[76] Raine CS. Membrane specialisations between demyelinated axons and astroglia in chronic EAE lesions and multiple sclerosis plaques. Nature. 1978;**275**(5678):326-327

[77] Reier PJ, Houle JD. The glial scar: Its bearing on axonal elongation and transplantation approaches to CNS repair. Advances in Neurology. 1988;**47**:87-138

[78] Eng LF, Reier PJ, Houle JD. Astrocyte activation and fibrous gliosis: Glial fibrillary acidic protein immunostaining of astrocytes following intraspinal cord grafting of fetal CNS tissue. Progress in Brain Research. 1987;**71**:439-455

[79] Holley JE et al. Astrocyte characterization in the multiple sclerosis glial scar. Neuropathology and Applied Neurobiology. 2003;**29**(5):434-444

[80] Smith ME, Eng LF. Glial fibrillary acidic protein in chronic relapsing experimental allergic encephalomyelitis in SJL/J mice. Journal of Neuroscience Research. 1987;**18**(1):203-208

[81] Franklin RJ, Gilson JM, Blakemore WF. Local recruitment of remyelinating cells in the repair of demyelination in the central nervous system. Journal of Neuroscience Research. 1997;**50**(2):337-344

[82] Bannerman P et al. Astrogliosis in EAE spinal cord: Derivation from radial glia, and relationships to oligodendroglia. Glia. 2007;**55**(1):57-64

[83] Sloka JS, Stefanelli M. The mechanism of action of methylprednisolone in the treatment of multiple sclerosis. Multiple Sclerosis. 2005;**11**(4):425-432

[84] Revel M. Interferon-beta in the treatment of relapsing-remitting multiple sclerosis. Pharmacology & Therapeutics. 2003;**100**(1):49-62

[85] Berger JR, Koralnik IJ. Progressive multifocal leukoencephalopathy and natalizumab—Unforeseen consequences. The New England Journal of Medicine. 2005;**353**(4):414-416

[86] Izrael M et al. Astrocytes in pathogenesis of ALS disease and potential translation into clinic. In: Gentile MT, D'Amato LC, editors. Astrocyte - Physiology and Pathology. London: IntechOpen; 2018. pp. 93-118. ISBN 978-953-51-5760-1

[87] Link H, Huang YM. Oligoclonal bands in multiple sclerosis cerebrospinal fluid: An update on methodology and clinical usefulness. Journal of Neuroimmunology. 2006;**180**(1-2):17-28

[88] Salou M et al. Expanded CD8 T-cell sharing between periphery and CNS in multiple sclerosis. Annals of Clinical Translational Neurology. 2015;**2**(6):609-622

[89] Sawcer S et al. Genetic risk and a primary role for cell-mediated immune mechanisms in multiple sclerosis. Nature. 2011;**476**(7359):214-219

[90] Parnell GP, Booth DR. The multiple sclerosis (MS) genetic risk factors indicate both acquired and innate

immune cell subsets contribute to MS pathogenesis and identify novel therapeutic opportunities. Frontiers in Immunology. 2017;**8**:425

[91] Junker A et al. Multiple sclerosis: T-cell receptor expression in distinct brain regions. Brain. 2007;**130**(Pt 11):2789-2799

[92] Alves Sousa AP et al. Comprehensive analysis of TCR-beta repertoire in patients with neurological immune-mediated disorders. Scientific Reports. 2019;**9**(1):344

[93] Kitze B et al. Myelin-specific T lymphocytes in multiple sclerosis patients and healthy individuals. Journal of Neuroimmunology. 1988;**20**(2-3):237

[94] Lourbopoulos A et al. Cyclization of PLP139-151 peptide reduces its encephalitogenic potential in experimental autoimmune encephalomyelitis. Bioorganic & Medicinal Chemistry. 2018;**26**(9):2221-2228

[95] Arneth B. Early activation of CD4+ and CD8+ T lymphocytes by myelin basic protein in subjects with MS. Journal of Translational Medicine. 2015;**13**:341

[96] Cohen JA et al. Autologous hematopoietic cell transplantation for treatment-refractory relapsing multiple sclerosis: Position statement from the american society for blood and marrow transplantation. Biology of Blood and Marrow Transplantation. May 2019;**25**(5):845-854

[97] Alexander T et al. Hematopoietic stem cell therapy for autoimmune diseases: Clinical experience and mechanisms. Journal of Autoimmunity. 2018;**92**:35-46

[98] Fassas A et al. Peripheral blood stem cell transplantation in the treatment of progressive multiple sclerosis: First results of a pilot study. Bone Marrow Transplantation. 1997;**20**(8):631-638

[99] Burt RK et al. Effect of Nonmyeloablative hematopoietic stem cell transplantation vs continued disease-modifying therapy on disease progression in patients with relapsing-remitting multiple sclerosis: A randomized clinical trial. JAMA. 2019;**321**(2):165-174

[100] Atkins HL et al. Immunoablation and autologous haemopoietic stem-cell transplantation for aggressive multiple sclerosis: A multicentre single-group phase 2 trial. Lancet. 2016;**388**(10044):576-585

[101] Burman J et al. Autologous haematopoietic stem cell transplantation for aggressive multiple sclerosis: The Swedish experience. Journal of Neurology, Neurosurgery, and Psychiatry. 2014;**85**(10):1116-1121

[102] Chen B et al. Long-term efficacy of autologous haematopoietic stem cell transplantation in multiple sclerosis at a single institution in China. Neurological Sciences. 2012;**33**(4):881-886

[103] Moore JJ et al. Prospective phase II clinical trial of autologous haematopoietic stem cell transplant for treatment refractory multiple sclerosis. Journal of Neurology, Neurosurgery, and Psychiatry. 2019;**90**(5):514-521

[104] Curro D, Mancardi G. Autologous hematopoietic stem cell transplantation in multiple sclerosis: 20 years of experience. Neurological Sciences. 2016;**37**(6):857-865

[105] Muraro PA et al. Autologous haematopoietic stem cell transplantation for treatment of multiple sclerosis. Nature Reviews. Neurology. 2017;**13**(7):391-405

[106] Ankrum JA, Ong JF, Karp JM. Mesenchymal stem cells:

Immune evasive, not immune privileged. Nature Biotechnology. 2014;**32**(3):252-260

[107] Mahla RS. Stem cells applications in regenerative medicine and disease therapeutics. International Journal of Cell Biology. 2016;**2016**:6940283

[108] Dominici M et al. Minimal criteria for defining multipotent mesenchymal stromal cells. The International Society for Cellular Therapy position statement. Cytotherapy. 2006;**8**(4):315-317

[109] Abbasi-Kangevari M et al. Potential therapeutic features of human amniotic Mesenchymal stem cells in multiple sclerosis: Immunomodulation, inflammation suppression, angiogenesis promotion, oxidative stress inhibition, neurogenesis induction, MMPs regulation, and remyelination stimulation. Frontiers in Immunology. 2019;**10**:238

[110] Gregory CA, Prockop DJ, Spees JL. Non-hematopoietic bone marrow stem cells: Molecular control of expansion and differentiation. Experimental Cell Research. 2005;**306**(2):330-335

[111] Zhang X et al. Transplantation of autologous adipose stem cells lacks therapeutic efficacy in the experimental autoimmune encephalomyelitis model. PLoS One. 2014;**9**(1):e85007

[112] Wilkins A et al. Human bone marrow-derived mesenchymal stem cells secrete brain-derived neurotrophic factor which promotes neuronal survival *in vitro*. Stem Cell Research. 2009;**3**(1):63-70

[113] Zhang J et al. Human bone marrow stromal cell treatment improves neurological functional recovery in EAE mice. Experimental Neurology. 2005;**195**(1):16-26

[114] Bai L et al. Human bone marrow-derived mesenchymal stem cells induce Th2-polarized immune response and promote endogenous repair in animal models of multiple sclerosis. Glia. 2009;**57**(11):1192-1203

[115] Kassis I et al. Neuroprotection and immunomodulation with mesenchymal stem cells in chronic experimental autoimmune encephalomyelitis. Archives of Neurology. 2008;**65**(6):753-761

[116] Gordon D et al. Human mesenchymal stem cells abrogate experimental allergic encephalomyelitis after intraperitoneal injection, and with sparse CNS infiltration. Neuroscience Letters. 2008;**448**(1):71-73

[117] Mohyeddin Bonab M et al. Does mesenchymal stem cell therapy help multiple sclerosis patients? Report of a pilot study. Iranian Journal of Immunology. 2007;**4**(1):50-57

[118] Harris VK, Sadiq SA. Stem cell therapy in multiple sclerosis: A future perspective. Neurodegenerative Disease Management. 2015;**5**(3):167-170

[119] Harris VK et al. Clinical and pathological effects of intrathecal injection of mesenchymal stem cell-derived neural progenitors in an experimental model of multiple sclerosis. Journal of the Neurological Sciences. 2012;**313**(1-2):167-177

[120] Harris VK, Vyshkina T, Sadiq SA. Clinical safety of intrathecal administration of mesenchymal stromal cell-derived neural progenitors in multiple sclerosis. Cytotherapy. 2016;**18**(12):1476-1482

[121] Harris VK et al. Phase I trial of intrathecal mesenchymal stem cell-derived neural progenitors in progressive multiple sclerosis. EBioMedicine. 2018;**29**:23-30

[122] Scolding NJ et al. Cell-based therapeutic strategies for multiple sclerosis. Brain. 2017;**140**(11):2776-2796

[123] Temple S, Alvarez-Buylla A. Stem cells in the adult mammalian central nervous system. Current Opinion in Neurobiology. 1999;**9**(1):135-141

[124] Blakemore WF, Keirstead HS. The origin of remyelinating cells in the central nervous system. Journal of Neuroimmunology. 1999;**98**(1):69-76

[125] Ming GL, Song H. Adult neurogenesis in the mammalian brain: Significant answers and significant questions. Neuron. 2011;**70**(4):687-702

[126] Ben-Hur T et al. Transplanted multipotential neural precursor cells migrate into the inflamed white matter in response to experimental autoimmune encephalomyelitis. Glia. 2003;**41**(1):73-80

[127] Einstein O et al. Neural precursors attenuate autoimmune encephalomyelitis by peripheral immunosuppression. Annals of Neurology. 2007;**61**(3):209-218

[128] Zhang C et al. Treatment of multiple sclerosis by transplantation of neural stem cells derived from induced pluripotent stem cells. Science China. Life Sciences. 2016;**59**(9):950-957

[129] Izrael M et al. Human oligodendrocytes derived from embryonic stem cells: Effect of noggin on phenotypic differentiation *in vitro* and on myelination *in vivo*. Molecular and Cellular Neurosciences. 2007;**34**(3):310-323

[130] Einstein O et al. Transplanted neural precursor cells reduce brain inflammation to attenuate chronic experimental autoimmune encephalomyelitis. Experimental Neurology. 2006;**198**(2):275-284

[131] Goldman SA, Windrem MS. Cell replacement therapy in neurological disease. Philosophical Transactions of the Royal Society of London. Series B, Biological Sciences. 2006;**361**(1473):1463-1475

[132] Windrem MS et al. Fetal and adult human oligodendrocyte progenitor cell isolates myelinate the congenitally dysmyelinated brain. Nature Medicine. 2004;**10**(1):93-97

[133] Windrem MS et al. Neonatal chimerization with human glial progenitor cells can both remyelinate and rescue the otherwise lethally hypomyelinated shiverer mouse. Cell Stem Cell. 2008;**2**(6):553-565

[134] Sim FJ et al. CD140a identifies a population of highly myelinogenic, migration-competent and efficiently engrafting human oligodendrocyte progenitor cells. Nature Biotechnology. 2011;**29**(10):934-941

[135] Goldman SA, Nedergaard M, Windrem MS. Glial progenitor cell-based treatment and modeling of neurological disease. Science. 2012;**338**(6106):491-495

[136] Thiruvalluvan A et al. Survival and functionality of human induced pluripotent stem cell-derived oligodendrocytes in a nonhuman primate model for multiple sclerosis. Stem Cells Translational Medicine. 2016;**5**(11):1550-1561

[137] Lepore AC et al. Human glial-restricted progenitor transplantation into cervical spinal cord of the SOD1 mouse model of ALS. PLoS One. 2011;**6**(10):e25968

[138] Izrael M et al. Safety and efficacy of human embryonic stem cell-derived astrocytes following intrathecal transplantation in SOD1(G93A) and NSG animal models. Stem Cell Research & Therapy. 2018;**9**(1):152

[139] Davies JE et al. Transplanted astrocytes derived from BMP- or CNTF-treated glial-restricted precursors

have opposite effects on recovery and allodynia after spinal cord injury. Journal of Biology. 2008;7(7):24

[140] Davies SJ et al. Transplantation of specific human astrocytes promotes functional recovery after spinal cord injury. PLoS One. 2011;**6**(3):e17328

[141] Noble M et al. Precursor cell biology and the development of astrocyte transplantation therapies: Lessons from spinal cord injury. Neurotherapeutics. 2011;**8**(4):677-693

[142] Jiang P et al. Human iPSC-derived immature astroglia promote oligodendrogenesis by increasing TIMP-1 secretion. Cell Reports. 2016;**15**(6):1303-1315

[143] Jiang P et al. hESC-derived Olig2+ progenitors generate a subtype of astroglia with protective effects against ischaemic brain injury. Nature Communications. 2013;**4**:2196

[144] Haas C, Fischer I. Human astrocytes derived from glial restricted progenitors support regeneration of the injured spinal cord. Journal of Neurotrauma. 2013;**30**(12):1035-1052

[145] Chandrasekaran A et al. Astrocyte differentiation of human pluripotent stem cells: New tools for neurological disorder research. Frontiers in Cellular Neuroscience. 2016;**10**:215

[146] Ciani L, Salinas PC. WNTs in the vertebrate nervous system: From patterning to neuronal connectivity. Nature Reviews. Neuroscience. 2005;**6**(5):351-362

[147] Krencik R et al. Specification of transplantable astroglial subtypes from human pluripotent stem cells. Nature Biotechnology. 2011;**29**(6):528-534

[148] Krencik R, Zhang SC. Directed differentiation of functional astroglial subtypes from human pluripotent stem cells. Nature Protocols. 2011;**6**(11):1710-1717

[149] Tcw J et al. An efficient platform for astrocyte differentiation from human induced pluripotent stem cells. Stem Cell Reports. 2017;**9**(2):600-614

[150] Zhang PW et al. Generation of GFAP::GFP astrocyte reporter lines from human adult fibroblast-derived iPS cells using zinc-finger nuclease technology. Glia. 2016;**64**(1):63-75

[151] Peljto M, Wichterle H. Programming embryonic stem cells to neuronal subtypes. Current Opinion in Neurobiology. 2011;**21**(1):43-51

[152] Caiazzo M et al. Direct conversion of fibroblasts into functional astrocytes by defined transcription factors. Stem Cell Reports. 2015;**4**(1):25-36

[153] Tian E et al. Small-molecule-based lineage reprogramming creates functional astrocytes. Cell Reports. 2016;**16**(3):781-792